Studies in Logic

Volume 27

Inconsistent Geometry

Volume 18
Classification Theory for Abstract Elementary Classes
Saharon Shelah

Volume 19
The Foundations of Mathematics
Kenneth Kunen

Volume 20
Classification Theory for Abstract Elementary Classes, Volume 2
Saharon Shelah

Volume 21
The Many Sides of Logic
Walter Carnielli, Marcelo E. Coniglio, Itala M. Loffredo D'Ottaviano, eds.

Volume 22
The Axiom of Choice
John L. Bell

Volume 23
The Logic of Fiction
John Woods, with a Foreword by Nicholas Griffin

Volume 24
Studies in Diagrammatology and Diagram Praxis
Olga Pombo and Alexander Gerner

Volume 25
The Analytical Way: Proceedings of the 6th European Congress of Analytical Philosophy
Tadeusz Czarnecki, Katarzyna Kijania-Placek, Olga Poller and Jan Woleński, eds.

Volume 26
Philosophical Aspects of Symbolic Reasoning in Early Modern Mathematics
Albrecht Heeffer and Maarten Van Dyck, eds.

Volume 27
Inconsistent Geometry
Chris Mortensen

Studies in Logic Series Editor
Dov Gabbay   dov.gabbay@kcl.ac.uk

Inconsistent Geometry

Chris Mortensen

© Individual author and College Publications 2010.
All rights reserved.

ISBN 978-1-84890-022-6

College Publications
Scientific Director: Dov Gabbay
Managing Director: Jane Spurr
Department of Computer Science
King's College London, Strand, London WC2R 2LS, UK

www.collegepublications.co.uk

Original cover design by orchid creative   www.orchidcreative.co.uk
Printed by Lightning Source, Milton Keynes, UK

All rights reserved. No part of this publication may be reproduced, stored in a retrieval system or transmitted in any form, or by any means, electronic, mechanical, photocopying, recording or otherwise without prior permission, in writing, from the publisher.

Contents

Contents		v
Preface		viii
I	**Paraconsistent Logics and Geometrical Theories**	**1**
1	**Paraconsistent Logic and Topology**	**3**
1.1	Introduction	3
1.2	Natural Logic is Paraconsistent	5
2	**Separation Principles and Identity**	**11**
2.1	Introduction	11
2.2	Separation Principles	13
2.3	Conditions on Identity and Disidentity	16
3	**Group Logics**	**19**
3.1	Introduction	19
3.2	Łukasiewicz Logic	20
3.3	Extending the Value Space	24
3.4	Abelian Logic	27
4	**Ulam Games**	**31**
4.1	Introduction	31
4.2	Ulam Games	31
4.3	Two Example Games with One Lie	32
4.4	Logics	35
4.5	The General Case $L > 1$	36
4.6	Paraconsistency	38
5	**Symmetry**	**41**
5.1	The Idea of Symmetry	41
5.2	An Inconsistent Approach ...	42

	5.3	A Formal Theory	43
	5.4	Recovering the Consistent Theory	46
	5.5	Symmetries of Reflection	48
	5.6	The Dihedral Group	48

6 Homomorphisms 51
  6.1 General Algebras 51
  6.2 Groups 53

7 Homology 57
  7.1 Introduction 57
  7.2 Simplices 57
  7.3 The Routley Functor on Groups 61
  7.4 An Application to Homology 63

II Inconsistent Images 67

8 Impossible Pictures 69
  8.1 Introduction 69
  8.2 Consistent Mathematical Approaches 71
  8.3 Conclusion 74

9 Logical Analysis of Necker Cubes 77
  9.1 Introduction 77
  9.2 Level 1 Analysis: Two Dimensions 80
  9.3 Level 2 Analysis: Three Dimensions 84
  9.4 Level 3: Faces 86

10 Linear Algebra and Necker Cubes 89
  10.1 Introduction 89
  10.2 Primary and Secondary Matrices 89
  10.3 Determinants 92
  10.4 Nullity 92
  10.5 The Structure of the Null Space 93
  10.6 The Unit Equation 95
  10.7 Conclusion 96

11 Linear Algebra & the Routley Functor 98
  11.1 Introduction 98
  11.2 Elementary Operations on Matrices 99
  11.3 The Routley Star on Matrices 100
  11.4 Transposes 103
  11.5 Order, Covariance and Contravariance 104
  11.6 Conclusion 104

CONTENTS

12 Necker Chains   **106**
12.1 Introduction 106
12.2 Chained Neckers 107
12.3 Degree of Inconsistency 110
12.4 Elementary Row and Column Operations 112
12.5 The Routley Functor 114
12.6 Order, Covariance and Contravariance 115
12.7 Conclusion 115

13 The Triangle   **117**
13.1 Introduction 117
13.2 The Triangle is a Paradox 118
13.3 The Triangle is an Occlusion Paradox 119
13.4 From Cognitive Science to Logic 123
13.5 From Logic to Mathematics 125
13.6 Conclusion 127

14 The Stairs   **129**
14.1 Introduction 129
14.2 Logic 130
14.3 Cognition 131
14.4 Mathematics 131
14.5 Conclusion 134

15 The Fork   **135**
15.1 Introduction 135
15.2 Logic 136
15.3 Cognition 137
15.4 Mathematics 138
15.5 Conclusion 140

Bibliography   **142**

A Gallery of Inconsistent Images   **146**

B Mostly B&W Images   **156**

Index   **160**

Preface

This book is a sequel to *Inconsistent Mathematics* (Kluwer 1995). That work was somewhat neglectful of the huge topic of geometry, I must confess. This was in part because at the time I was labouring under a misapprehension, namely that to produce inconsistent mathematical theories based on figures in continuous space, it was necessary that the functional properties of the real numbers be drastically truncated to avoid trivialising, and it was not apparent how to do this satisfactorily. However, subsequently it became clear to me that this puts the cart before the horse, and that geometrical structures provide modellings for inconsistent theories directly, bringing their functional properties along with them, as it were. Exploring such modellings is the task of the first part of this book.

At the same time, I was uncomfortably aware that inconsistent mathematics had not addressed the so-called impossible pictures, or inconsistent images, drawn by Reutersvaard, Escher, the Penroses and many others. Yet inconsistent mathematics would seem to be exactly the right tool for this job. Classical consistent mathematics has made only a handful of contributions on the subject, and in a sense cannot hope to deal with it adequately. Inconsistent objects do not exist in the external world, they are products of our cognition. So it would seem that we must have inconsistent mathematical theories to describe the contents of our cognitive states produced by these images, the sense that "I see it but it cannot exist". But classical consistent mathematics does not treat such theories.

This book takes up both of these themes, the former in Part 1, the latter in Part 2. It is not a general textbook on geometry, it makes no claims to exhaustiveness. It takes up various geometrical theories that have interesting inconsistent aspects, such as topology, separation principles, symmetry, homology, Ulam games, and of course group theory; before moving to impossible images. Readers should forgive me if I omit their favourite geometrical topics, and hopefully will be inspired to ex-

PREFACE

tend the project in new directions. I aim to steer a course through the common ground between mathematics and logic, to inform geometers about the applications of their subject to the theory of inconsistency, and to inform logicians about the rich array of sources available to them from within geometry.

I wish to thank Steve Leishman and Peter Quigley for their assistance, particularly with the graphics. Leishman is also a co-author of Chapter 10, and Quigley is a co-author of Chapter 4.

Coloured images throughout the text, or in the Appendix, are offered as a gallery of what is possible for inconsistent images. The text does not generally address these coloured images directly. All these images including the Appendix were created by Leishman, with the exceptions: Figure 2.2 p18, Figure 13.12 p130, Figure 15.3 p143 (all CM), and Figure 6.1 p56 and small modifications elsewhere (PQ). Colour by PQ and SL, captions by CM. Many more inconsistent images and discussions can be found at the website:

http://www.hss.adelaide.edu.au/philosophy/inconsistent-images/

Thanks are due to Bruno Ernst for permission to reproduce Figure 8.1, namely Oscar Reutersvaard's *Opus 1*, which can be seen in Ernst's *The Eye Beguiled* (1986, 69).

Finally I wish to thank my wife Cathy for her encouragement and support while this was being written.

Chris Mortensen
Dept of Philosophy
The University of Adelaide
25 April, 2010

Part I

Paraconsistent Logics and Geometrical Theories

Chapter 1

Paraconsistent Logic and Topology

1.1 Introduction

In this book, I will explore the connections between paraconsistent logic and geometry. For the reader unfamiliar with the concept of paraconsistency, let it be said that paraconsistent logics are those which tolerate inconsistency without deductive collapse. As such, they enable us to explore the rich structures that are to be found within the inconsistent. To deal with inconsistent theories, all paraconsistent logics reject the rule *Ex Contradictione Quodlibet* (ECQ for short). This is the rule that *from contradictory premisses any conclusion may be validly deduced*. In symbols, ECQ is $A, \neg A \vdash B$. Classical two-valued logic claims that ECQ is valid, thus paraconsistent logics all reject that much of classical two-valued logic.

In a slogan, the claim that ECQ is valid amounts to the thesis that *the inconsistent has no structure*. If ECQ were a valid rule of deduction, then rigorous deductive exploration of any structure within inconsistent theories would be impossible. To the extent that humans display the capacity for contradiction-toleration, humans are paraconsistent reasoners, and therefore two-valued logic is an erroneous representation of their reasoning capacities. A huge number of different paraconsistent logics have been discovered over the last 40 years or so, which exploit various stategies for contradiction-containment.

However, this is not primarily a book about different sorts of paraconsistent logics. It is, rather, an application of the techniques of the theory of inconsistency, to the case of geometry. In *Inconsistent Mathe-*

matics (1995), geometry was not to the forefront; and so it is necessary to rectify that relative neglect of a large area of mathematics. It also became clear in that book that inconsistent theories tend to be invariant over many different (paraconsistent) background logics. This means that a small number of simple paraconsistent logics can be assumed, so that the real mathematical work can proceed unhindered (though we will also see that there is interaction between mathematical structures and different logics for negation).

Logicians have often steered clear of the subject of geometry. One can speculate as to reasons. Lattices appear widely in the semantics of logics, yet lattices have only fairly rudimentary geometrical representations (Hasse diagrams). Other geometrical structures do not obviously lend themselves to interpretations for logics (though we will go some way here to improving on the situation). To date, theories within inconsistent mathematics have tended to focus on algebraic number theory to the exclusion of geometry. In the end, perhaps it is simply that logicians, despite the warning over Plato's Academy (*Do Not Enter Here Unless You Know Geometry*), have not been particularly well trained in the difficult subject of geometry.

This book will go about the project of geometrising logic in two broad ways. In the first part, the aim is to find models for paraconsistent logics exploiting geometrical objects, and conversely to show how logics and logical theories develop out of classes of geometrical objects.

In the second part of the book, we will tackle the topic of providing mathematical descriptions of the so-called impossible pictures, drawn by Reutersvaard, Escher, and others. Only a few classical mathematicians have made an attempt at this difficult class of problems; for example, Penrose, Cowan, Francis. But invariably, their otherwise adroit accounts miss something essential. Being classical and thus consistent, they cannot give a sense of the content of the experience of the impossible, *I see it but it can't exist*. The only way to complete the story, I propose, is to give an inconsistent theory as the content of the experience. It will stand to the experience of the impossible in somewhat the way that projective geometry stands to the experience of perspective, only the theory will perforce be inconsistent.

The importance of inconsistent images is enormous, I think. Even sceptics who disbelieve in paraconsistency have difficulty in insisting that the inconsistent has no structure, when confronted with these examples. In turn, success in our endeavour will strengthen the case for paraconsistency, by displaying the content of human geometrical experiences and reasoning as inconsistent. It must be stressed that this sort of motivation for the theory of inconsistency is *epistemic* or *cognitive* in character.

That is, the study of inconsistency is seen to arise out of the presence of inconsistencies within the contents of our cognitive structures. This is sometimes called *weak* paraconsistency. It contrasts with *strong* paraconsistency, also called *dialetheism,* which is the claim that some contradictions are true, true in the world not just true in thought. In practice, there is no weakness in weak paraconsistency, it permits study of exactly the same structures, and by exactly the same methods. It is simply that the cognitive motivation for weak paraconsistency is more general and less testing than the ontological case for dialetheism, and can be accepted by all. This is not to say that I reject dialetheism, only that this book does not depend on it. Having said that, to the extent that mathematics contains propositions which are true but contradictory, then this book demonstrates dialetheism about *mathematical* truth.

1.2 Natural Logic is Paraconsistent

The core of the theory of inconsistency is the concept of negation: a theory is inconsistent if it contains at least one pair of statements of the form $A, \neg A$. Other logical operations get into the act only to the extent that they interact with negation (which of course they do). As a preliminary, then, in this section we show that there is at least one natural paraconsistent negation, namely *closed set negation*. This requires us first to rehearse some known facts about modal logic.

Consider first the formal semantics of propositional modal logic. Modal logic adds to the usual Boolean operators $\{\&, \vee, \neg\}$, the unary propositional operator $\Box$, where $\Box A$ is interpreted as "it is necessary that A". Additional Boolean operators such as $\supset$ and $\equiv$ are defined in the usual way, and the modal operator $\Diamond$, interpreted as "it is possible that" is defined as $\neg\Box\neg$. The possible worlds semantics for modal logic constructs models using a set W of possible worlds. Propositions hold at some worlds and do not hold at others, so we can write "Aw" for "the proposition A holds at world w". For example, if A is the proposition that snow is white, then Aw is the proposition that snow is white holds in world w. This is given an algebraic setting by associating with each proposition A, a set $[A]$ of members of W, to be interpreted as the set of worlds at which A holds. Thus we can define Aw to mean $w \in [A]$. The simplest set of conditions governing the behaviour of the operators $\{\&, \vee, \neg, \Box\}$ is the following:

(1) $[A]$ is a subset of W, for all propositions A.

(2) $[\neg A] = -[A]$, the set complement of $[A]$.

(3) $[A \& B] = [A] \cap [B]$

(4) $[A \vee B] = [A] \cup [B]$

(5) $[\Box A] = W$ if $[A] = W$, else $[\Box A] = \{\}$, the null set.

It is then simple to prove that:

(6) $[\Diamond A] = W$ if $[A] \neq \{\}$, else $[\Diamond A] = \{\}$.

We can define the *deducibility relation* $\models$ *in a model* $[\ ]$ by: $A \models B$ iff $[A] \subseteq [B]$. Then define a proposition to be a *semantic theorem* just in case it is true in all worlds in all models, that is $[A] = W$ in all models. The set of semantic theorems so obtained coincides exactly with the set of provable theorems of the logic S5. The simplicity of this semantics has convinced the large majority of modal logicians that S5 is the preferred modal logic as a description of the properties of necessity and possibility.

The idea that W might have a topological structure (W, O) with open sets O allows a generalisation, which gives us our first geometrical connection. Thus we can replace the semantic condition (5) for $[\Box A]$ by the condition:

(5.1) $[\Box A] = int[A]$, the interior of $[A]$.

With this condition, we find instead that the semantic theorems coincide exactly with the provable theorems of S4. Furthermore, it is apparent that we can readily recover the S5 case with the additional stipulation that the topology O on W is the the indiscrete topology. Note that the change to (5.1) implies that the semantical condition for possibility, condition (6), changes to:

(6.1) $[\Diamond A] = cl[A]$, the closure of $[A]$.

Modal logic is not the only place where logic connects with topology. One can link the behaviour of $\{\&, \vee, \neg\}$, particularly $\neg$, with the topological structure on W. Thus, instead of (1) above, one stipulates that *the semantic value of any proposition shall be an open set*. This is not so arbitrary as it might seem at first glance. The set W of possible worlds can be construed as a phase space of possible configurations of some system, and it seems that it is a contingent possibility that propositions might be true only on open sets of points. As a special case, for example, W might be an index of times, so that processes are described by propositions changing over time, and processes might well only ever take place on open sets of times.

1.2. NATURAL LOGIC IS PARACONSISTENT

The stipulation that propositions are only ever true on open sets of points requires that the operators $\{\&, \vee, \neg\}$ shall be functional on the open sets. This is no problem for $\{\&, \vee\}$, since open sets are closed w.r.t. (finite) intersections and (arbitrary) unions. However, for $\neg$ to be an operator on open sets, it cannot generally be the set (Boolean) complement. Instead, one associates it with the *open complement*, that it, the largest open set contained in the set complement, which might be identical with the set complement but need not be. Algebras of open sets are called *Heyting algebras*. Summarising, we take just (1)–(4), and replace (1) and (2) by

(1.1) $[A] \in O$, the open sets on W.

(2.1) $[\neg A]$ = the open complement of $[A]$.

There is also a natural implication operator $\Rightarrow$, which is not definable in terms of $\{\&, \vee, \neg\}$, but is definable semantically as:

(7) $[A \Rightarrow B] = [A] \Rightarrow [B] = \cup \{O : O \text{ is open and } [A] \cap O \subseteq [B]\}$.

As before, a semantic theorem is any formula which holds at all points of W in all models. This gives the theorems of *intuitionist logic*. Logics arising from open set topological spaces are referred to as *open set logics*, *OSL* for short.

It is well known that the above semantic features mean that intuitionist logic supports *incomplete* theories, that is, theories in which neither A nor $\neg A$ holds for at least one proposition A. To see this, consider a proposition A and let w be any point on $bdry[A]$, the boundary of $[A]$. Now $[A]$ is an open set, so w is not in $[A]$, hence we do not have Aw. But neither do we have $\neg Aw$, since that requires $w \in [\neg A]$ = the open complement of $[A]$, which is disjoint from $bdry[A]$. Again, neither A nor $\neg A$ hold at w, so the theory containing the propositions holding at w, is incomplete. So one can describe w, viewed as a possible world, as an *incomplete world*.

Now, the topological duality between open and closed, is mirrored in a duality between intuitionist logic and (one variant of) paraconsistent logic. To recall for the reader, a theory is *trivial* if it contains every proposition (of the language in question), and thus a logic is paraconsistent if it has at least one non-trivial inconsistent theory. Now, instead of stipulating that propositions hold only on open sets of points, we stipulate that propositions hold only on closed sets of points. As before, this seems like a condition which could conceivably hold on some phase space. But then, in order to have negation as a natural operation, we must identify the negation of a proposition A as holding on the *closed complement* of $[A]$, which is the smallest closed set containing the set complement. That is, replace (1) and (2) instead by:

(1.1) $A \in C$, the closed sets on W.

(2.1) $[\neg A]$ = the closed complement of $[A]$.

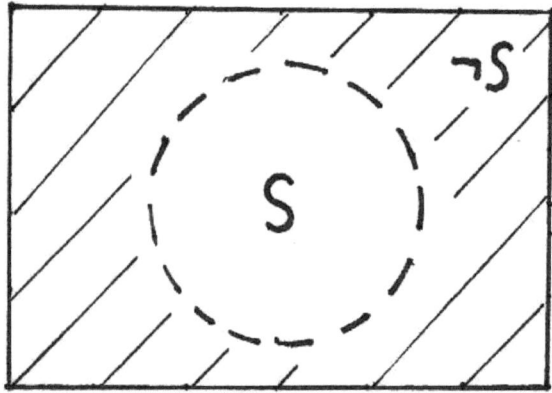

Figure 1.1: Open set complement

We also have, instead of a condition for implication (7), a natural *logical subtraction* operation, given by:

(7.1) $[A - B] = [A] - [B] = \cap \{C : C \text{ is closed and } [A] \subseteq [B] \cup C\}$.

Algebras of closed sets are called *Brouwerian algebras*.

This change requires re-thinking the semantic condition governing theoremhood. In moving from open sets to closed sets, one is, in effect, reversing the order on the lattice. That is, the bijection that turns open sets into their (closed) set complements, is contravariant with respect to subset inclusion, which is the natural order on the set lattice. Now, in any lattice-theoretic value space of more than two values, there is in general more than one (non-trivial) filter on the space. Membership of the filter then serves to determine membership of a theory. That is, a given Bouwerian lattice of closed sets, or for that matter a given Heyting lattice of open sets, can support more than one theory for the same value function [ ], determined by the different filters on the lattice. However, the most natural dual of the Heyting condition (that a theorem is determined by the property of holding at every point in W), is the

1.2. NATURAL LOGIC IS PARACONSISTENT

condition of holding at *some* point in W. The set of propositions that hold at some point in all closed set models is called *closed set logic*, and logics arising from particular closed set topological spaces are called *closed set logics*, *CSL* for short. The set of theorems of closed set logic in the connectives $\{\&, \vee, \neg\}$ is exactly those of classical logic. However, the deductive relationship is weaker than classical logic.

To see this, let $[A]$ be a closed set and let w be any point on $bdry[A]$. Since $w \in [A]$, we have Aw. But also, since $w \in [\neg A]$ = the closed complement of $[A]$, we also have $\neg Aw$. That is, the theory consisting of the propositions holding at w, is inconsistent. The world w can thus be described as an *inconsistent world*. The theory is not generally trivial, however, since many propositions may fail to hold at w. In this sense, closed set logic is paraconsistent, that is, it supports non-trivial inconsistent theories.

One more point is about implication. The properties of the natural dual of intuitionist implication $\Rightarrow$, namely logical subtraction $-$, prevent it from being a reasonable implication. For example $A - A$ fails to be a theorem. This is not particularly paradoxical, since dually intuitionist logic lacks a natural subtraction, which Boolean logic contains. Furthermore, intuitionist $\Rightarrow$ is not the only reasonable implication around. For instance, on any (bounded) lattice whatsoever, there is always at least one reasonable implication, namely that operator $A \Rightarrow B$ which equals W if $[A] \subseteq [B]$, else equals $\{\}$.

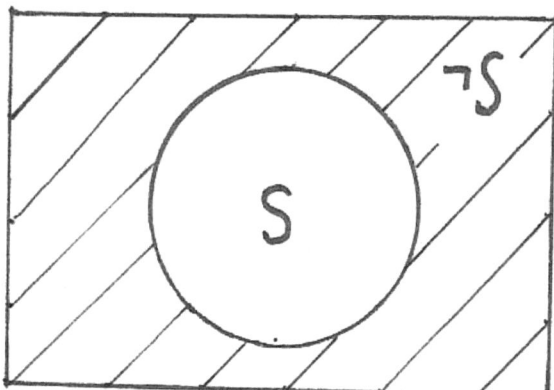

Figure 1.2: Closed set complement

The simplest forms of these logics arise from the topology on the real numbers R, with three open sets $\{R, R - \{0\}, \{\}\}$, and three closed sets $\{R, \{0\}, \{\}\}$. These give rise to two three-valued logics, respectively open set and closed set, as follows.

&	1	2	3	∨	1	2	3	¬(OSL)	¬(CSL)
1	1	2	3		1	1	1	3	3
2	2	2	3		1	2	2	3	1
3	3	3	3		1	2	3	1	1

We recall that the designated value for OSL is the single value R, which is represented by the top value 1. On the other hand, for CSL the designated values are the two upper values 1 and 2, that is everything but the bottom value 3 representing $\{\}$. Hence in CSL, a formula such as $A\&\neg A$ takes the designated value 2 when A is 2, whereas $\neg\neg(A\&\neg A)$ takes the undesignated value 3. This shows, again, that contradictions do not trivialise. In terms of the closed sets themselves, when $[A] = \{0\}$, then $[A\&\neg A] = \{0\}$, which is not a subset of $[\neg\neg(A\&\neg A)] = \{\}$, hence the latter formula is not deducible from the former, so CSL is paraconsistent.

Finally, we note that the most general logic ought to be able to deal with both inconsistency and incompleteness. Ignoring implication and pseudo-complement, the best way to achieve this is to take the cross product of the above two three-valued logics. This simplifies to the four natural values, $\{R, \{0\}, R - \{0\}, \{\}\}$. Conjunction and disjunction are as before set intersection and union. When it comes to negation, there is only one way to make it functional on this value space, and that is to associate it with the *topological complement*, which is to say the open complement of an open set and the closed complement of a closed set. In the case of a clopen set, these coincide with one another and with set complement. That is, we replace (2) with:

(2.3) $[\neg A]$ = the topological complement of $[A]$.

An interesting consequence of this definition is that we can have $[A] \subseteq [B]$ without $[\neg B] \subseteq [\neg A]$. For example, if $[A]$ is open and $[B]$ its closure, then $[\neg B]$ includes their common boundary, but $[\neg A]$ does not as it is open. This is a kind of limited failure of contraposition which is significant because it occurs in the context of a well-motivated semantics.

*

In this chapter, we looked at a first example of the connection between geometry and logic, particularly paraconsistent logic. We saw that a natural account of paraconsistent negation arises. In the next chapter we continue the theme of topological negation, by showing that separation principles in topological spaces are reflected in the containment properties of inconsistent logical theories.

Chapter 2

Separation Principles and Identity

2.1 Introduction

In this chapter, we continue with the theme of topological spaces and their distinctive concept of topological negation, which, as we saw in the first chapter, admits both inconsistent and incomplete theories. It will be seen that there is an interaction between the properties of such theories, and various *topological separation* conditions.

We want to construct theories on a topological space (X, O, C). with open sets O and closed sets C. (In classical mathematics, this is usually shortened to (X, O), or even X; but our purposes are served by noticing the closed sets as well.) The **language** of these theories will consist in *subsets* of the following. *Terms*: names $\{a, b, ...\}$ for all members of X, names $\{S, T, S_1...\}$ for all subsets of X, including the null set $\{\}$. For convenience we can take them as naming themselves. *Binary Predicates, or Relations*: $\{=, \in\}$. *Unary* predicates are obtained by filling one of the spaces in a binary predicate, e.g. " $= a$", "$a =$", "$\in S$". *Atomic sentences*: of the form $a = b$, and $a \in S$. *Logical operators*: $\&, \vee, \neg$. Implication and quantifiers are omitted as unnecessary at this stage, but the language evidently has a predicate structure.

We want first to define suitable models on topological spaces, and then theories on those models. For the results of this section, we consider only unary predicates built from $\in$, using the same free variable x, ranging over individuals. This suffices to provide modellings for the binary predicates, and permits logical compounding of the extensions of unary predicates. In the next section, we study identity.

A *model* is a function which assigns to each individual name a, an element $[a]$ of X, and to each unary predicate F of the language a subset $[F]$ of X. Atomic predicates having the same free variable compound under the operators to give composite extensions: $[F\&G] = [F] \cap [G]$, $[F \vee G] = [F] \cup [G]$.

For negation, a *generalised* topological complement is useful. We stipulate that:

Definition 1 (Topological complement)

(1) $\neg[F] := (bdry[F] \cap [F]) \cup int\overline{[F]}$.

This amounts to augmenting the interior of the set complement of $[F]$ with those parts of the boundary which are part of $[F]$. When $[F]$ is open or closed, this coincides with the topological complement of the previous chapter, and in particular when $[F]$ is clopen, it is the set complement, since the boundary of a clopen set is null. When $[F]$ is neither open nor closed, $[\neg F]$ includes just those boundary points which are also part of $[F]$. Topological negation is then given by the obvious $[\neg F] = \neg[F]$.

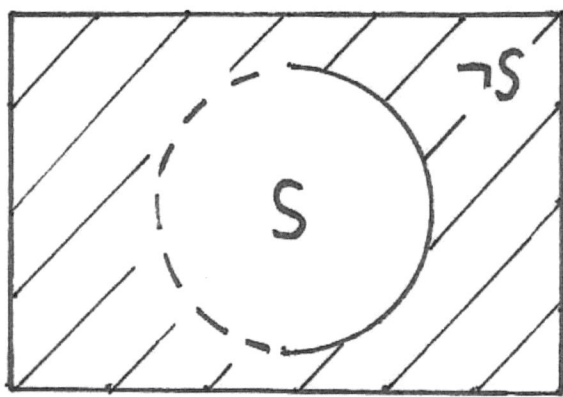

Figure 2.1: General topological complement

Sentences are formed by substituting an appropriate name for the free variable. The *theory of* the model $[\,]$ *on* the space (X, O, C) is then generated by stipulating that the sentence Fa *holds* in the theory on the model $[\,]$, or F *holds at* a in $[\,]$, iff $[a] \in [F]$. We recall from the first chapter that the extensions of such predicates, being subsets of a space with a topology, such as a set of worlds might have, can also be identified as propositions, and Fa interpreted as "the proposition F holds at world a". Note that the condition for negation implies that if $a \in bdry[F] \cap [F]$,

2.2. SEPARATION PRINCIPLES

then both Fa and $\neg Fa$ hold, even if $[F]$ is neither open nor closed; while if $a \in bdry[F]$ but $a \notin [F]$, then neither Fa nor $\neg Fa$ hold. That is, theories arising from sets which are neither open nor closed, are both inconsistent and incomplete.

These preliminaries enable us to investigate separation principles.

2.2 Separation Principles

Separation principles for topological spaces are so called because they stipulate conditions on the open or closed sets by means of which distinct but nearby points are separated or excluded. So we need some definitions. In increasing order of strength:

Definition 2 *(Separation principles)*

(1) A topological space (X, O, C) is a T_1 space iff for any pair of distinct points, each has a neighbourhood which excludes the other.

*(2) (X, O, C) is a T_2 space (or a **Hausdorff space**) iff every distinct pair of points are contained in a pair of disjoint open sets.*

*(3) (X, O, C) is **normal** iff every disjoint pair of closed sets are contained in a pair of disjoint open sets.*

*(4) (X, O, C) has the **discrete** topology iff every set is open (and therefore also closed).*

See Kelley (1955, 112). Note that the definition of T_1 is equivalent to: every singleton is closed. (Simmons 1963, 130). Now we have a simple result.

Theorem 3 *If (X, O, C) has the discrete topology, then every theory on it is both consistent and complete.*

Proof. Let F be any predicate of whatever logical complexity, having the single free variable x. Since the topology is discrete, $[F]$ is clopen. Hence $[\neg F]$ is the set complement of $[F]$. Thus not both Fa and $\neg Fa$ hold, for any a in X. Thus the sentences of the theory are consistent. And if Fa does not hold, then a is not in $[F]$, so it is in $[\neg F]$, so that $\neg Fa$ holds. That is completeness also. ∎

For an appropriate converse:

14   CHAPTER 2. SEPARATION PRINCIPLES AND IDENTITY

Theorem 4 *If (X, O, C) is a T_1 space and every theory on it is consistent, then the space has the discrete topology.*

Proof. If (X, O, C) is T_1 and is not discrete, then we show how to make a theory which is inconsistent. Since it is T_1, every singleton is closed. If all singletons were open, then any subset would be open, since open sets are closed under arbitrary unions. So some singleton $\{a\}$ is closed and not open. Let the model be $[\in \{a\}] = \{a\}$. Clearly, $a \in \{a\}$ holds. But $\{a\}$ is closed and not open, so $\neg[\in \{a\}] = X$. This means that $\neg a \in \{a\}$ also holds and the theory is inconsistent. ∎

Notice that for any b distinct from a, the sentence $b \in \{a\}$ does not hold, so the theory is not trivial.

Theorem 5 *(X, O, C) has the discrete topology iff (X, O, C) is T_1 and every theory on it is consistent.*

Proof. Follows from the previous two theorems together with the fact that the discrete topology is T_1. ∎

These theorems between them furnish us with examples of inconsistent and incomplete theories on topological spaces. For inconsistent theories, one cannot allow the discrete topology. But any non-discrete T_1 space will yield them. If $\{a\}$ is any closed non-open singleton, take any theory in which $a \in \{a\}$ holds, or more generally any theory in which $[F] = \{a\}$. These theories are in general non-trivial, for example if they contain a name b for some distinct element, then Fb does not hold.

For incomplete theories, we must have a space which is not T_1. A simple case is the 3 member set $X = \{a, b, c\}$, with $O = \{X, \{a, b\}, \{a, c\}, \{a\}, \{\}\}$, and $C = \{X, \{b, c\}, \{b\}, \{c\}, \{\}\}$. Clearly, $b \in \{a\}$ does not hold. But also, $\neg[\in \{a\}] = \{\}$, so that $\neg b \in \{a\}$ also does not hold. The theory is therefore incomplete. At the same time, this theory is inconsistent, since both $b \in \{b\}$ and $\neg b \in \{b\}$ hold.

To deal with stronger separation conditions, we need further definitions:

Definition 6 *If S is any subset of the space X, then we will say that the **membership-theory** of S, or $Th(\in S)$, is the theory on X with names for all members of X, satisfying $[a] = \{a\}$, plus a predicate "$\in \Sigma$" with $[\in \Sigma] = S$.*

Definition 7 *A membership-theory on X is a **one-point theory** iff it is the membership-theory of some singleton $\{a\}$ where $a \in X$.*

2.2. SEPARATION PRINCIPLES

Definition 8 *One theory is an **atomic extension** of another iff the atomic sentences of the latter are a subset of the former. Two theories are **atomically disjoint** if their atomic sentences are disjoint.*

Now we have two results which characterise T_2 and normal spaces respectively.

Theorem 9 *The space X is Hausdorff iff every pair of one-point membership-theories of subsets of X have a pair of consistent membership-theories which are atomically disjoint atomic extensions.*

Proof. R to L: Suppose the RHS and let $a, b \in X$ for distinct a, b. It has to be shown that a, b can be separated by disjoint open sets. The membership theories $Th(\in \{a\})$ and $Th(\in \{b\})$ are 1-point, and so by the RHS, they have disjoint consistent atomic extensions. These are membership-theories, so let S, T be the sets of which they are the membership-theories. Since $a \in \Sigma$ is an atomic sentence in $Th(\in A)$, it is also in $Th(\in S)$, so that $a \in S$. Similarly $b \in T$. If S and T were not disjoint, say c were in both of them, then $c \in \Sigma$ would be in both $Th(\in S)$ and $Th(\in T)$ and these theories would not be atomically disjoint. Since they are disjoint, so must S and T be. Finally, S and T must be open sets, for if they contained even one point of their boundaries, say $a \in S$, then by the modelling condition for negation, $\neg a \in S$ holds, contradicting the consistency of S.

L to R. Conversely, suppose the space is Hausdorff, and let $Th(\in \{a\})$, $Th(\in \{b\})$ be arbitrary 1-point membership theories. From the Hausdorff condition, a and b are contained in disjoint open sets S, T. Since S, T are open, they lack all their boundary points, so their membership theories are consistent. The sole atomic sentence of $Th(\in \{a\})$ is $a \in \Sigma$. But also a is in S, so $a \in \Sigma$ holds in $Th(\in S)$ as well; hence the latter membership-theory is an atomic extension of the former. Similarly for b and T. Finally, the membership-theories of S, T are atomically disjoint; for if they were not, say they shared $c \in \Sigma$, then c would be in both S and T, contradicting their disjointness. ∎

Theorem 10 *The space X is normal iff every pair of atomically disjoint complete membership-theories of subsets of X have a pair of consistent membership-theories which are atomically disjoint atomic extensions.*

Proof. R to L. Suppose the RHS and let U, V be a disjoint pair of closed sets. It has to be shown that they are contained in a disjoint pair of open sets. The membership-theories $Th(\in U)$, $Th(\in V)$ are atomically disjoint, and so by the RHS there is a pair of atomically disjoint membership-theories $Th(\in S)$, $Th(\in T)$ that are atomic extensions. If $a \in U$, then $a \in \Sigma$ is an atomic sentence of $Th(\in U)$, so it is an atomic sentence of $Th(\in S)$, so that $a \in S$. That is, $U \subseteq S$. Similarly, $V \subseteq T$. Now S and T are disjoint, for otherwise if they had a common member c, then $c \in \Sigma$ would be a common member of $Th(\in S)$ and $Th(\in T)$, contradicting their disjointness. Finally, the latter two theories are consistent, so that their corresponding sets S, T must be open, for if even one member b of their boundaries were included in one of them, then both $b \in \Sigma$ and $\neg b \in \Sigma$ would hold, contradicting consistency.

L to R. Conversely, suppose that X is normal, and let $Th(\in U)$, $Th(\in V)$ be atomically disjoint complete membership-theories with U, $V \subseteq X$. By normality, there are disjoint open sets S, T with $U \subseteq S$, $V \subseteq T$ respectively. Since S, T are open, their membership-theories are consistent. If these membership-theories were not atomically disjoint, then they would both contain some $c \in \Sigma$, contradicting the disjointness of S and T. Finally, since $U \subseteq S$, $Th(\in S)$ is an atomic extension of $Th(\in U)$, and correspondingly for V and T. ∎

These results are somewhat prolix, due to their being esentially a restatement of the separation axioms in the language of logical theories and operations. Nonetheless, they have a character of their own, since they exploit the properties of topological negation to give extendability of the atomic parts of pairs of complete possibly inconsistent theories, to consistent theories containing them.

Having considered the interaction of membership-theories and separation conditions, we now proceed to explore modellings for identity on topological spaces. We see that, in a similar fashion, topological negation plays a significant role.

2.3 Conditions on Identity and Disidentity

We move to consider identity and disidentity. In the setting of monadic predicates, it is natural to model $[= a]$ and $[a =]$ as $\{a\}$. That is:

(2) $[= a] = [a =] = \{a\}$, all a in X

In spaces with as weak a topology as T_1, if negation is interpreted as topological negation, then this implies that $\neg a = a$ holds for every a. This is because all disidentities hold, since $\neg[= a] = X$. There are several ways to escape this consequence. The simplest way to escape, is to vary the conditions for a negated identity predicate so as to make it classical:

(3) $[\neg = a] = [\neg a =] = \overline{\{a\}}$, the set complement of $\{a\}$.

2.3. CONDITIONS ON IDENTITY AND DISIDENTITY

But there are other ways to model identity. Thus one can replace (2) by:

(2.1) $[= a] = [a =] = \{a\} \cap S$, for some set $S \subseteq X$, and all $a \in X$.

It follows from (2.1) that if $a \in S$ then $[= a] = \{a\}$, else $[= a] = \{\}$. This has the effect that only members of S are self-identical; that is if $b \notin S$ then $b = b$ does not hold. If we retain (3) for negation, then the theory of identity remains consistent and complete, albeit having some non-self-identical elements $\neg b = b$ for those $b \notin S$.

If, on the other hand, instead of (3) we use topological negation (1), then if X is a T_1 space then $\neg a = a$ holds for all a. That is, the contents of S are characterised by inconsistent identity, while outside S everything is non-self-identical. As S contracts, the region where no identities hold expands, so this can be dubbed the "scorched earth model".

Alternatively, one can replace (2.1) with:

(2.2) $[= a] = [a =] = \{a\} \cap (S \cup \neg S)$

If S is closed, then the set $S \cup \neg S$ is the whole space X, so that $a = a$ holds for every a. In this case, the topological negation of $\{a\} \cap (S \cup \neg S)$ is X, so that $\neg a = a$ holds for every a.

Otherwise, if S is not closed, then if a is a boundary point of S not in S, (that is $a \in \overline{S} \cap bdry(S)$), then $a = a$ does not hold. Moreover, $a \notin int[F]$, so that $\neg a = a$ fails also.

There is also:

(2.3) if $a \in S$ then $[= a] = [a =] = S$, else $[= a] = [a =] = \{a\}$.

This has the effect that the members of S are identified with one another. The methodology of identifying elements is ubiquitous in topology. We will use this in later chapters. This type of model can also be used to study the functional properties of inconsistent theories as determined by quotient topologies arising from various equivalence relations (Mortensen 1995, Ch9).

There are also other options for modelling the negation of identity sentences. For example, one could have:

(3.1) $\neg [= a] = \overline{\{a\}} \cup (\{a\} \cap bdry(S) \cap S)$.

This has the effect that $\neg b = a$ holds for every b other than a; but also $\neg a = a$ for every a on the boundary of S as long as it is included in S. For example, if S is closed and not open, then it is surrounded by a boundary of contradictory self-identies. This can be called the "ring of fire" model.

*

In this chapter we have seen that there are significant interactions between the topological properties of spaces, and the logical properties of theories on them. In a later chapter we will turn to a different account of boundary, to give different insights. But now we must take account of the concept of a *group*. Group theory is the ubiquitous tool of the geometer, so any work on geometry must address them. On the other hand, logicians have not been much into group theory. But if geometry and logic are linked, then there must be some connections. Now there have been some connections made by logicians, and so in the next chapter we look at these. This will give an alternative conception of logics, and particularly negation.

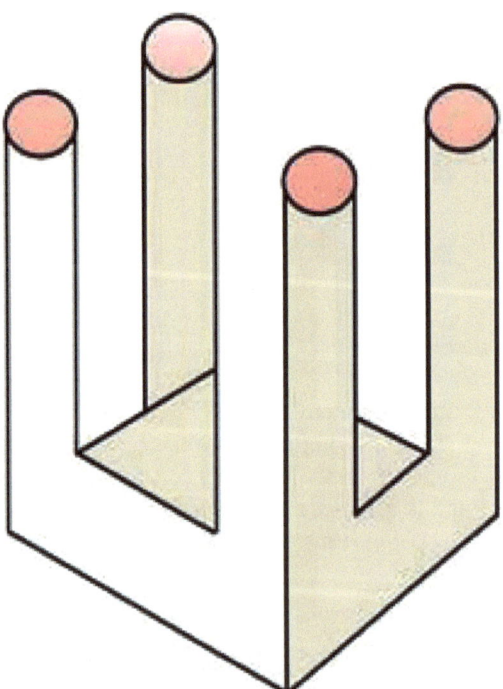

Figure 2.2: Four-poster

Chapter 3

Group Logics

3.1 Introduction

Until the second half of the 19th century, the emphasis in geometry had been on synthetic geometry, that is setting up axioms and proving theorems from them, in the way that Euclid did. In 1872 in Erlangen, Felix Klein made a break with the past, by enunciating what came to be called the Erlangen Program. This was the call to describe geometries and geometrical objects by means of groups, specifically groups of transformations that leave geometrical objects and properties invariant. Klein's influence was immense, and today the use of group theory in geometry is ubiquitous. In turn, this gives us a link between logic and geometry, because there also exists a significant body of literature by logicians, on group logics. To recall for the reader:

Definition 11 (*Groups*) *A group is a set equipped with three items: (1) a binary operation $\circ$ which is associative, that is $x \circ (y \circ z) = (x \circ y) \circ z$, (2) a neutral or null or identity element id, with the property that $x \circ id = id \circ x = x$, and (3) for each element x a unique inverse $-x$, with the property that $x \circ -x = -x \circ x = id$. If in addition the group satisfies $x \circ y = y \circ x$ for every x and y, the group is said to be Abelian.*

Examples of groups are the integers, rationals and reals with the group operation addition $+$, $id = 0$, and additive inverse $-$. When a group arises from an addition or addition-like operation, it is called an *additive* group, its operation is written as an addition, and the group identity is written 0. Other examples of groups are the positive rationals and positive reals, with the group operation multiplication $\times$, $id = 1$, and multiplicative inverse $^{-1}$. When a group arises from a multiplication or multiplication-like operation, it is called a *multiplicative* group, and its operations are written as multiplication and multiplicative inverse respectively, and the group identity is written 1.

The basic intuitive understanding of the group operations calls out for logical application. First, the idea that the elements of a group can be *transformations*, that is *actions* of a certain sort, lends itself to the thought that a group is a space of propositions, namely the propositions made true by the carrying out of the actions. This follows from the well-known result *Cayley's Theorem*, which says that any abstract group is isomorphic with a group of transformations. Second, the basic group operation for combining operations is readily understood as "and then", which is undoubtedly a kind of conjunction. Importantly, it is a kind of conjunction which is *associative but not necessarily commutative*. Indeed, it is commutative iff the group is Abelian, and many interesting examples of non-Abelian groups are known. Third, the group inverse is an obvious candidate for a kind of negation, one which obeys the law of Double Negation, as Boolean negation does. The role of the group identity is more indirect, which we see as we go along.

From just this much, it is fair to say that *group theory is its own logic*. Groups are in a reasonable sense made up of propositions, and group theory sets itself to describe its subject matter sufficiently with the tools that it postulates, tools which can be viewed as logic-like operations on spaces of propositions. Logicians, it might be thought, have nothing more to add to the topic.

But of course, logic aims to study operations beyond this sparse set of tools. What of disjunction and implication, for example? Without these, the expressive power of group logic cannot be regarded as very great. The question thus arises of whether we can construct such further operations from within group theory. There have been a number of approaches to this problem, and in this chapter they will be surveyed.

3.2 Łukasiewicz Logic

The basic tools of group theory outlined above do not provide us with very much to go on with. However, if we add the stipulation that the group is *lattice ordered*, then there are rich resources to be exploited. Łukasiewicz logic is a well-known logic which, as we see in this section, arises in a natural way from lattice ordered groups. The groups we deal with, will generally be additive groups, and we adjust our symbolism accordingly. These results are basically due to Chang (1958). For a thorough discussion of the topics of this section, see Cignoli *et al,* (2000)

3.2. ŁUKASIEWICZ LOGIC

Definition 12 *A lattice is a set equipped with a partial order $\leq$ (that is, reflexive, antisymmetric and transitive), and having the property that for any pair of elements the least upper bound and greatest lower bound of those elements exists. When the least upper bound and greatest lower bound of all (infinite) sets of elements exist, the lattice is said to be complete as a lattice.*

Definition 13 *A lattice ordered group (or ℓ-group) is an Abelian group with a partial order that respects the group operation, that is, for any x, y, z in the group: if $x \leq y$ then $z + x \leq z + y$, and equivalently $x + z \leq y + z$.*

The additive groups of reals, rationals and integers, with their natural orders, are all examples of ℓ-groups, as are the multiplicative groups of positive reals and rationals. Being unbounded, they are not complete as lattices.

The construction of Łukasiewicz logic from ℓ-groups proceeds by way of an intermediate concept, that of an *MV-algebra*, which provides the algebraic semantics for the logic.

Definition 14 *An MV-algebra is an algebra $\langle A, \oplus, \neg, 0 \rangle$, with a binary operation $\oplus$, a unary operation $\neg$, and a constant 0 satisfying the following equations:*

MV1: $x \oplus (y \oplus z) = (x \oplus y) \oplus z$

MV2: $x \oplus y = y \oplus x$

MV3: $x \oplus 0 = x$

MV4: $\neg\neg x = x$

MV5: $x \oplus \neg 0 = \neg 0$

MV6: $\neg(\neg x \oplus y) \oplus y = \neg(\neg y \oplus x) \oplus x$.

To define MV-algebras within ℓ-groups, the lattice order needs to be bounded. This is necessary for two reasons: first a top value is needed to be the designated value for Łukasiewicz logic, and second the order needs upper and lower bounds to be the values of quantifiers. One way to achieve this is to choose an arbitrary element u of the ℓ-group, called a *strong unit*, and take this to be the negation of zero in MV5 above. Essentially this means that we take our algebra to be defined on the closed interval $[0, u]$ bounded by 0 and u. For example, a simple way to achieve this outcome is to begin with the ℓ-group of reals under addition, and utilise 1 as the strong unit. Then we can write:

Definition 15 *(MV-algebras from ℓ–groups)*

$x \oplus y := glb(1, x + y)$ *(fission)*
$\neg x := 1 - x$ *(negation)*

Given that the group we start with is the reals under addition, glb is of course minimum, though the former is more general. Now we have the following result, which establishes the construction of MV-algebras from ℓ–groups. Let G be the ℓ–group of reals under addition, let $[0, 1]$ be the closed interval of reals bounded by 0 and 1, let $\oplus, \neg$ be as just defined, and let the structure $\langle [0, 1], \oplus, \neg, 0 \rangle$ be denoted by $\mathbf{\Gamma}(G, 1)$.

Theorem 16 $\mathbf{\Gamma}(G, 1)$ *is an MV-algebra.*

Proof. See Cignoli (2000, 2.1.2). ∎

MV-algebras permit the definition of algebraic counterparts of many common logical operators. For example, in line with the terminology of Meyer-Slaney (2002, see later in this chapter), we can distinguish between *extensional* structure (disjunction, conjunction and the quantifiers), and *intensional* structure (fission, fusion, implication, negation). While these two structures interact, the former are essentially determined by the lattice order, and the latter by the group structure. Thus we have:

Definition 17 *(MV-algebra operations)*

(i) $x \vee y := lub(x, y)$ *(disjunction)*

(ii) $x \wedge y := glb(x, y)$ *(conjunction)*

(iii) $x \odot y := \neg(\neg x \oplus \neg y)$, *(equivalently $:= 0 \vee (x + y - 1)$) (fusion)*

(iv) $x \to y := \neg x \oplus y$, *(equivalently $:= 1 \wedge (1 - x + y)$) (implication)*

(v) $x \ominus y := x \odot \neg y$, *(equivalently $:= 0 \vee (x - y)$) (subtraction).*

In light of (i), the previous definition of fission is equivalently $x \oplus y = 1 \wedge (x + y)$.

Now MV-algebras are characteristic for *infinite-valued Łukasiewicz logic L_∞*. This is a logic with reasonable properties, and can be axiomatised in the $(\to \neg, \vee, \wedge)$ language as follows, defining $A \leftrightarrow B$ as $(A \to B) \wedge (B \to A)$:

Definition 18 *(Axioms and rules for L_∞)*

L1: $(A \to B) \to ((B \to C) \to (A \to C))$

3.2. ŁUKASIEWICZ LOGIC

L2: $A \to (B \to A)$

L3: $(A \to \neg B) \to (B \to \neg A)$

L4: $((A \to B) \to B) \to ((B \to A) \to A)$

L5: $((A \to B) \to B) \leftrightarrow (A \vee B)$

L6: $(A \wedge B) \leftrightarrow \neg(\neg A \vee \neg B)$.

Rule: $A, A \to B \vdash B$ *(modus ponens).*

To show that every theorem of $Ł_\infty$ is valid in the above MV-algebra requires the stipulation of a designated value or values in the latter. The value 1 is taken as the sole designated value. Then:

Theorem 19 *Any theorem of $Ł_\infty$ is valid in the MV-algebra $\Gamma(G, 1)$.*

Proof. It is a straightforward argument to show that every axiom takes the value 1 in any assignment of values from $[0, 1]$, using the definitions of the MV-operations above, and that the modus ponens rule preserves this property. ∎

In addition to infinite-valued Łukasiewicz logic, there are finite-valued Łukasiewicz logics $Ł_n$ (with more theorems). These are defined by restricting the value interval $[0, 1]$ to the n rational values:

$$0, 1/(n-1), 2/(n-1), ..., (n-2)/(n-1), 1$$

This restriction respects the above operations, so we have a construction of finite-valued Łukasiewicz logics from ℓ-groups and MV-algebras. All Łukasiewicz logics are sublogics of classical logic, and so their intuitive motivations are no worse than those that classical logic has seemed to have. None are paraconsistent.

To recover MV-algebras from $Ł_\infty$, it suffices to apply the usual construction of a Lindenbaum algebra. Begin with the set of formulae $\mathcal{F}$ of $Ł_\infty$, and define on it the equivalence relation $\equiv$ by: $A \equiv B := (\vdash A \to B$ and $\vdash B \to A)$. It is straightforward to show that $\equiv$ so defined respects the basic operators; that is, it is a congruence on the formula algebra. This enables the set of the operators necessary for an MV-algebra, namely $\{\oplus, \neg, 0\}$, to be well-defined on the set of equivalence classes $\mathcal{F}/\equiv$.

So we have:

Theorem 20 $\langle \mathcal{F}/\equiv, \oplus, \neg, 0 \rangle$ *is an MV-algebra.*

Figure 3.1: Hasse diagram of n-valued Łukasiewicz logic

Proof. See Cignoli (2000, 4.4). ∎

Finally, to complete the circle, ℓ–groups are recoverable from MV-algebras. This is a somewhat more complex argument, and we will not go into it here; though it should be registered that the natural order is definable by: $x \leq y$ iff $x \to y = 1$. It follows from this result that ℓ–groups can be recovered from Ł_∞, that is that group theory can be done in Łukasiewicz logic. It also follows that ℓ–groups and MV-algebras are equivalent as categories. For details, see Cignoli (2000, Chaps 2-4).

Łukasiewicz logic has a number of applications. Aristotle thought that the *law of excluded middle LEM:* $A \vee \neg A$ should fail in certain cases, such as contingent statements about the future. LEM can be seen to fail in Łukasiewicz logic if we assign A the value 0.5. Another application, of a geometrical kind, that we will see in a later chapter is Ulam games. For our purposes at this point, however, it suffices to note that these are natural logics that arise from group theory, and thus, indirectly, from geometry.

3.3 Extending the Value Space

There are some less-than-optimal features of the above algebras. First, Łukasiewicz logics are not paraconsistent. Second, negation in these algebras is not quite the group inverse. Third, the operation fission is not quite the group operation ∘. In this section, we show how to remedy the first two of these features in a reasonable way. We consider the third in a later section.

3.3. EXTENDING THE VALUE SPACE

Chang (1963) saw that negation could be made to coincide with group inverse by extending the elements of the algebra, and so the value space for the corresponding logic, to the closed interval of reals $[-1, +1]$. On this set, Chang defined MV^* algebras:

Definition 21 (*MV* algebras*)
$x \oplus y := 1 \wedge (-1 \vee (x + y))$
$\neg x := -x.$

The first of these evidently derives from the group operation by bounding above and below. The second is exactly the coincidence of negation and inverse.

This construction in turn gives rise, in a similar way to the previous section, to a logic Chang called Ł*. This has the following definition:

Definition 22 (*Axioms and rules for Ł**)
Abbreviations:
$A^+ := (A \to 1) \to 1$
$A^- := (A \to \neg 1) \to \neg 1$
$A \vee B := ((A^+ \to B^+)^+ \to (\neg A)^-) \to ((B^- \to A^-)^- \to A^-)$
$A \wedge B := \neg(\neg A \vee \neg B)$
$A \leftrightarrow B := (A \to B) \wedge (B \to A)$

Axioms:
Ł*1: $(A \to B) \leftrightarrow (\neg B \to \neg A)$
Ł*2: $A \leftrightarrow ((B \to B) \to A)$
Ł*3: $\neg(A \to B) \leftrightarrow (B \to A)$
Ł*4: $A \to 1$
Ł*5: $1 \leftrightarrow ((1 \to A) \to 1)$
Ł*6: $((A \to 1) \to ((B \to 1) \to C)) \to ((B \to 1) \to ((A \to 1) \to C))$
Ł*7: $(A \to B) \leftrightarrow ((B^+ \to A^-) \to (A^+ \to B^-))$
Ł*8: $(A \to (\neg A \to B))^+ \leftrightarrow (A^+ \to (\neg(A^+) \to B^+))$
Ł*9: $(A \to (B \vee C)) \leftrightarrow ((A \to B) \vee (A \to C)).$
Ł*10: $(A \vee (B \vee C)) \leftrightarrow ((A \vee B) \vee C)$

Rules:
Ł*R1: $A, A \to B \vdash B$
Ł*R2: $A \to B, C \to D \vdash (B \to C) \to (A \to D)$
Ł*R3: $A \vdash A^-.$

This is undoubtedly less simple that Ł$_\infty$. We note that some simplification is afforded by the following facts in the MV*-semantics:

Theorem 23

(a) $x \to y = 1 \wedge (-1 \vee (y - x))$

(b) $x^+ = 0 \vee x$

(c) $x^- = -(0 \vee -x)$

(d) If $x \to y = 0$, then $x = y$.

Proof. See e.g. Lewin-Sagastume (2002, 395). ∎

In particular, notice that x^- is not the negative of x, but is the same as x if x is negative, and otherwise is 0.

The important thing to note here is that in the MV*−algebraic sematics, the designated values are no longer the top element alone, but the stretch of positive reals $[0, 1]$. This is important because it ensures that the *the corresponding logic L^* is paraconsistent*. This follows from the well-known point that a logic whose semantics has a designated value which is a fixed-point with respect to negation, is always paraconsistent. Simply set $A = \neg A = 0$, and take B to be any formula sharing no variables with A and taking a negative value. Then the premises of ECQ all take a designated value, while its conclusion does not, so the conclusion of this instance of ECQ does not follow from its premises. On the other hand, L^* is *not a relevant logic*. Relevant logics, due to Anderson and Belnap (1975), are an important subset of the paraconsistent logics, having the *relevance property*: namely that *a conditional is a theorem only if the antecedent and consequent of the conditional share an atomic sentence letter*. Since it is an axiom of L^* that $A \leftrightarrow ((B \to B) \to A)$, then picking any theorem for A and any formula with no letters in common for B, we can detach A and thus get irrelevance.

Lewin and Sagastume axiomatise the part of L^* that takes the (designated) value 0, which they call L_0^*. This is given by:

Definition 24 *(Axioms and rules for L_0^*)*
Axioms: The axioms are those of L^*, except that L^*4 is dropped.
Rules: The first two rules are the same as for L^*, together with the rule:

$A \vdash B \to (A \to B)$.

This logic has various theorems which have seemed intuitively reasonable to many, such as $A \to A$, and including $(A \to A) \leftrightarrow (B \to B)$. However, the effect of axiomatising those formulae which take just the fixed-point value, is that a formula is a theorem iff its negation is. This is not a property that one would wish for the general class of logical truths. Of course, it is not absurd if one's aim is to provide (by axiomatic means) a class of formulae which hold iff their negations do, which one might wish to capture if one had independent interest in a class of formulae which was a boundary.

3.4 Abelian Logic

In (1989) and (2002), Meyer and Slaney extracted a logic they called **A** from the linearly ordered group of the integers **Z**. Now it might be argued that the integers are number-theoretic, and not directly a geometrical object; but still they are undoubtedly applicable to geometry.

Meyer and Slaney managed to satisfy all three of the desiderata nominated in the previous section: paraconsistency, negation as the group inverse, and fission as the group operation. But it came with a cost: the logic **A** had a number of interesting but unusual properties, such as one might think were not in accordance with orthodox intuitions of what should hold in logics. Even so, in light of the relative paucity of group-theoretic semantics for logics, it is striking that **A** is well-behaved in many respects.

To begin with, notice a general constraint on any attempt to satisfy these desiderata.

Theorem 25 *In any group, $-(-x + -y) = y + x$.*

Proof. First, note that $-y + -x$ is a group inverse of $x + y$, since $(-y + -x) + (x + y)$ computes to 0, applying associativity. But $-(x + y)$ is also a group inverse of $x + y$, and group inverses are unique, so $-(x + y) = -y + -x$. The theorem follows by substituting $-x, -y$ for x, y respectively, and cancellation of double negations. ∎

It follows that in any Abelian group, $-(-x + -y) = x + y$. In consequence, if as one of our desiderata we require fission to be the group operation, and we define, as is usual, fusion by means of the De Morgan equivalence, then fission and fusion are the same. Group logic cannot distinguish them. Meyer and Slaney point out this consequence (1989, 254). One thing that might be said in favour of it, is a point about fusion that has not come up yet. In the first section of this chapter, we noted that the group operation was a kind of ordered conjunction, generally non-Abelian. Fusion is intensional conjunction. Chang defines the group operation as intensional disjunction. But why not try the group operation as fusion? The above result shows that, other things being equal, it is possible to do so. Nonetheless, the point remains that fusion and fission still cannot be distinguished, whichever is preferred. Moreover, there is always the point that Chang's choice can be independently justified by the naturalness of the results.

Semantics for **A** in the additive group Z are as expected:

Definition 26 *(Semantical definitions for **A**)*

An interpretation $I : \mathcal{F} \longrightarrow Z$ satisfies:
S1: $I(t) := 0$
S2: $I(x \oplus y) := x + y$
S3: $I(x \to y) := y - x$
S4: $I(x \wedge y) := \min(x, y)$
S5: $I(x \vee y) := \max(x, y)$
S6: $I(\neg x) := -x$
S7: A formula A is valid on an interpretation I iff $I(A) \geq 0$.

A sound and complete axiomatisation for **A** is given by the following axioms and rules (using the nomenclature of Meyer-Slaney 2002):

Definition 27 *(Axioms and Rules for **A**)*

Axioms:
AxA: $((A \to B) \to B) \to A$
AxB: $(B \to C) \to ((A \to B) \to (A \to C))$
AxC: $(A \to (B \to C)) \to (B \to (A \to C))$
AxI: $A \to A$
Ax$\wedge$E: $(A \wedge B) \to A$, also $(A \wedge B) \to B$
Ax$\to\wedge$: $((A \to B) \wedge (A \to C)) \to (A \to (B \wedge C))$
Ax$\wedge\vee$: $(A \wedge (B \vee C)) \leftrightarrow ((A \wedge B) \vee (A \wedge C))$
Ax$\vee\to$: $((A \to C) \wedge (B \to C)) \leftrightarrow ((A \vee B) \to C)$
Ax$\oplus\to$: $((A \oplus B) \to C) \leftrightarrow (A \to (B \to C))$
Axt$\to$: $A \leftrightarrow (t \to A)$

Rules:
Ru$\to$E: $A \to B, A \vdash B$
Ru$\wedge$I: $A, B \vdash A \wedge B$.

A is not merely axiomatisable, but decidable, since the non-theorems can be enumerated by working through interpretations in the integers. Interestingly, **A** also has an "infinite model property", according to which *all non-trivial models of **A** are infinite*. It also shares with classical logic the maximality property of being *Post-complete;* that is, if any non-theorem is added and the result closed under the rules and uniform

3.4. ABELIAN LOGIC

substitution, then all formulae become provable. Finally, **A** is paraconsistent, for the same sorts of reasons as L* above (having a designated value which is a fixed point for negation, is sufficient for paraconsistency).

The above axioms and rules are all fairly uncontentious (or as uncontentious as anything can be in a disputed subject like logic), save for one. Attention is drawn to AxA, namely $((A \to B) \to B) \to A$. Meyer and Slaney call this the *axiom of relativity*. It is hard to marshal intuitions in support of this, and it is certainly not a theorem of classical logic, nor of the Łukasiewicz logics. They argue that it is an expression of double negation: the negation $\neg A$ of a formula A ought to be definable as $A \to f$, where f is defined as $\neg t$, and both are identified with 0. For then, substituting f for B in the axiom of relativity reduces it to $\neg\neg A \to A$, which is to say one half of the law of double negation. This holds by the group property $--A = A$.

*

This completes our survey of logics with group semantics. There is much more that could be shown about the structures in this chapter, but enough has been said for us to see that there are undeniable connections between group theory and logics, including paraconsistent logics. The group identity is an appropriate designated fixed point for negation, as long as negation is modelled by the group inverse. Even so, a lattice order is an appropriate addition to these structures, so as to have conjunction and disjunction with reasonable properties. It seems therefore that plausible logics cannot escape lattices somewhere in their semantics. Moreover, if a well-behaved paraconsistent negation (such as topological negation) can be defined on a lattice without recourse to group inverses, then there is the prospect of being able to do a lot of inconsistent mathematics independent of intensional concepts. This is not to say that group theory would be ignored. Rather, the inconsistent aspects of mathematics seem to be relatively invariant over large classes of logics with different intensional superstructures, so that they can be studied independently of intensional considerations.

In the next chapter, we turn to an application of Łukasiewicz logics, namely the theory of Ulam games. We will see that these are paraconsistent in a certain specific way, and have natural geometrical interpretations, particularly in higher dimensions.

Figure 3.2: Triangles

Chapter 4

Ulam Games

(co-authored with Peter Quigley)

4.1 Introduction

In this chapter, we further illustrate the connection between geometry and Łukasiewicz logics, by summarising and extending results about Ulam games. We will see that different classes of geometrical structures are reflected in different sizes of Łukasiewicz logics. Ulam games are in a *prima facie* sense paraconsistent; though this raises an interesting puzzle, in that Łukasiewicz logics are not. In the last section, we take up this question of how paraconsistency manifests itself in these games. To begin with, we must describe Ulam games.

4.2 Ulam Games

Given a finite set S of numbers (called the search space), and an unknown number $x \in S$, the task is to find x by a series of questions. A maximum number of lies L in the answers are permitted. It is allowed to have fewer than the maximum number in any game, but L must be known to all parties. The size of S is denoted by n, and for simplicity we can assume $S = \{1, 2, ..., n\}$. The case $L = 0$ of zero lies, is the familiar game of Twenty Questions. For example, if $L = 0, n = 10^6$, then x can be determined by a series of questions which halve the search space. The first question would be: does $x \in \{1, ..., 500,000\}$? After 20 questions, x is found. We consider Ulam games with L lies. The remarkable thing is that these are also decidable, and by means of Łukasiewicz logics. Until recently, the following facts were all that was known (see Cignoli 2000):

(1) Solving such games requires $Ł_{L+2}$, that is $(L+2)$-valued Łukasiewicz logics.

(2) The case $L = 1$ corresponds to 3-valued Łukasiewicz logic arising naturally on the vertices, edges and faces of n-dimensional cubes (n-cubes), where n is the size of the search space.

In a later paper (Mortensen-Quigley 2005), the geometrical modelling was extended to the general case $L > 1$. In this chapter we summarise these results. It will be seen that the number of lies makes a difference to the kinds of geometrical modellings given, which in turn reflects the size of the Łukasiewicz logic that characterises the problem.

4.3 Two Example Games with One Lie

We explain the algorithm for solving Ulam games by means of examples, beginning with one lie (3-valued logic).

Game I: $L = 1, S = \{1, 2\}$

There are two players, questioner and answerer. Both know that the number of lies ≤ 1. The number x is unknown to the questioner, who does know what the search space S is, and that its size is $n = 2$. We track the questioner's knowledge states $K0, K1, ...$ at times $t = 0, 1, ...$ At $t = 0$, the knowledge state is a vector of ones corresponding to the members of S, that is $K0 = (1, 1)$. At later times, the entries in Ki are values from $L + 2$-valued Łukasiewicz logic, that is in the present game of one lie, the logic is $Ł_3$, and the possible values for entries in Ki are $\{1, 1/2, 0\}$

Question 1: $x \in \{1\}$?
Answer: yes (may be a lie, in fact it is the truth)

Algorithm for Proceeding: An answer of *yes* is treated as follows. The entries in Ki for any number in any set to which the answer is *yes* stay the same. The entries for any number not in that set drop to the next lowest value of the Łukasiewicz logic being employed. That is, the entry drops by $1/(L+1)$. Thus, in the present game where $L = 1$, the entry for the number 2 drops to $1/2$. Thus the next knowledge state $K1 = (1, 1/2)$.

4.3. TWO EXAMPLE GAMES WITH ONE LIE

An answer of *no* is treated as an answer of *yes* to the opposite question, about the complement-set in S. This is equivalent to the logical operation of fusion $\odot$ of the previous knowledge-state point-by-point with 1 for those numbers which attracted the answer yes, and $L/(L+1)$ for those numbers which attracted the answer no. We can obviously form the latter sequence as a vector, and so we can say that $K(t+1) = Kt \odot$ (the vector of answers to Question $t+1$). Notice in passing how fusion functions appropriately as a kind of conjunction here.

Continuing on with Game I:
Question 2: $x \in \{2\}$?
Answer: yes (a lie)
Next knowledge state $K2 = (1/2, 1/2)$
Question 3: $x \in \{2\}$?
Answer: no
Next knowledge state $K3 = (1/2, 0)$

The game concludes when the knowledge state is a vector only one of whose entries is non-zero. Such knowledge states can be called *solution states*, and $K3$ is one such. The place of the non-zero entry corresponds to the previously unknown number, 1, which can be read off.

Game II: $L = 1, n = 3$

Let the unknown $x = 2$.
$K0 = (1, 1, 1)$
Question 1: $x \in \{1, 2\}$?
Answer: no (a lie)
$K1 = (1/2, 1/2, 1)$
Question 2: $x \in \{1, 3\}$?
Answer: no
$K2 = (0, 1/2, 1/2)$
Question 3: $x \in \{3\}$?
Answer: no
$K3 = (0, 1/2, 0)$ which is a solution state, and the solution is $x = 2$.

Geometric Representations

This procedure can be modelled isomorphically in a natural way by the vertices, edges and faces (generally n–faces) of the n–cube. Consider first Figure 4.1.

When $n = 2$, knowledge states correspond 1-1 to the vertices, edges and face of a square (2-cube). There are 4 vertices, 4 edges and 1 face. The

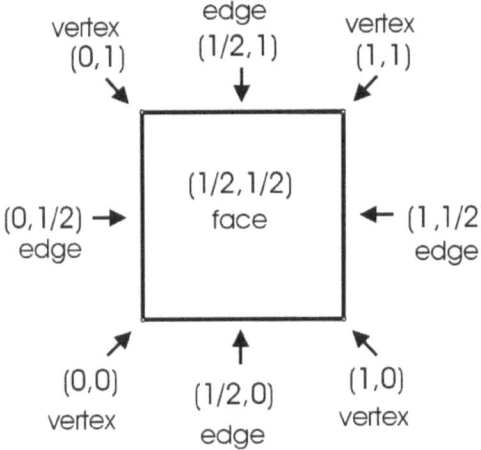

Figure 4.1: Game 1: $L = 1, n = 2$

coordinates of the vertices (0-faces) are (1,1), (1,0), (0,1) and (0,0). The latter value is never achieved in a Ulam game, since the game ends with one non-zero component of the coordinate, but serves as a base point. The edges (1-faces) are (1,1/2), (1/2,1), (1/2,0) and (0,1/2). The face (the whole 2-cube) is (1/2,1/2). The game begins at the point (1,1) and moves down the edges, perhaps occupying the whole face, until one of the solution states is reached. The four solution states are the two edges connected to (0,0), and their two non-zero endpoints.

Figure 4.2 illustrates Game II.

When $n = 3$, knowledge states correspond to the vertices, edges and faces of the 3-cube (including the whole cube itself). There are 8 vertices, corresponding to the 8 knowledge states (1,1,1), (1,1,0), (1,0,1), (0,1,1), (1,0,0), (0,1,0), (0,0,1), and (0,0,0). There are 12 edges, corresponding to states where just one component is 1/2 and the two others 0 or 1. The six faces correspond to the states where two components are 1/2 and the other component 0 or 1. The whole cube (3-face) is (1/2,1/2,1/2), which is a possible knowledge state. The 6 solution states are the 3 edges (0,0,1/2), (0,1/2,0) and (1/2,0,0) and their 3 nonzero endpoints (0,0,1), (0,1,0) and (1,0,0). It is apparent that the presence of a component 1/2 indicates a line in that dimension (uncertainty spread along that line, as it were), so that two components of 1/2 indicates a 2-face, and in general n components indicates an n−face.

4.4. LOGICS

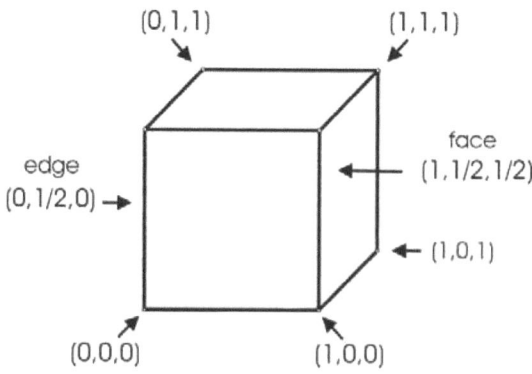

Figure 4.2: Game II: $L = 1, n = 3$

4.4 Logics

In passing, we note a further geometric structure here, what has been called *cubic logic* (see Rota-Metropolis 1978). The knowledge states corresponding to the faces of an n-cube can evidently be represented as functions $Kt \in \{1, 1/2, 0\}^{\{1,...,n\}}$, which in turn correspond 1-1 with the n-tuples of values $(Kt(1), Kt(2), ..., Kt(n))$. On the set of all such functions/knowledge states, there can be defined lattice operations pointwise:

Definition 28 *(Rota-Metropolis cubic logic)*

For each $j \in \{1, ..., n\}$,

(1) $(Kt1 \sqcup Kt2)(j) = Kt2(j)$ if $Kt1(j) = Kt2(j)$, otherwise $= 1/2$

(2) If $Kt1$ intersects $Kt2$, that is $\mid Kt1 - Kt2 \mid \leq 1/2$, then $(Kt1 \sqcap Kt2)(j) = Kt2(j)$ if $Kt1(j) = 1/2$, otherwise $= Kt1$

(3) If $Kt1 \sqcup Kt2 = Kt2$, then $(\Delta(Kt2, Kt1))(j) = 1 - Kt1(j)$ if $Kt2(j) = 1/2$, otherwise $= Kt1(j)$.

This is an upper semilattice, that is, least upper bound is well-defined, but greatest lower bound is not always defined. When $Kt2(j) = 1/2$, then $\Delta(Kt2, Kt1)(j) = $ the Lukasiewicz negation of $Kt1(j)$, otherwise Δ leaves $Kt1$ unchanged. For further discussion, see Mundici (2002, 398).

This is at most a partial logic. But there is a full modelling in Łukasiewicz logic on knowledge states. We have already seen Łukasiewicz fusion $\odot$ in use in the previous section, and we can define it (suppressing the component parameter j) by: $Kt1 \odot Kt2 := lub(0, Kt1+Kt2-1)$. The usual Łukasiewicz definitions of the other operators can then be applied: $\neg Kt := 1 - Kt$ (negation), $Kt1 \oplus Kt2 := \neg(\neg Kt1 \odot \neg Kt2)$ (fission), and $Kt1 \rightarrow Kt2 := \neg Kt1 \oplus Kt2$ (implication). As usual, negation has a fixed point, since the knowledge state $(1/2, ..., 1/2)$ is self-negating. The natural partial order is: $Kt1 \leq Kt2 := (all\ j)(Kt1(j) \leq Kt2(j))$. This can be understood (contravariantly) as: *Kt1 is at least as well informed as Kt2*, since the further we get down the tree of questions and answers, the further we get away from $(1, ..., 1)$ which is complete ignorance, and the closer we get to the solution states, which take the least values (apart from the null state, which is not a solution).

4.5 The General Case $L > 1$

In moving to the general case of L lies, we recall first that for $L = 1$ and general n, the corresponding figure is an n–cube. It is also known that for L lies, the corresponding Łukasiewicz logic is $(L+2)$-valued. It suffices, then, to produce geometrical figures that have the right number of values, $L + 2$, mapped onto them. This can readily be done by taking the 1-dimensional edges, and dividing each of them into $L+1$ equal portions. Thus the edge $(0, 0, ..., 1/2, ...0, 0)$ is broken at the points $(0, 0, ..., 1/(L+1), ..., 0, 0)$, ..., $(0, 0, ..., L/(L+1), ...0, 0)$. That is, what was the middle value of $1/2$ for the case $L = 1$ is split into L intermediate values $1/(L+1), ..., L/(L+1)$ in the general case.

Each of the knowledge states with only one non-zero component is a solution state, as before. Thus, since there are n components, and there are $L+1$ possible non-zero entries for each component, there are $n(L+1)$ distinct solution states.

This process can be likened to drawing lines across the 2–faces of the n–cube to join up the partitioning points (see diagrams below). However, there is one point to note. In the case $L = 1$, each vertex, edge, face and the whole n–cube corresponds to a knowledge state. But for general L this is not so, the internal partitioning lines do not correspond to knowledge states. They are the boundaries between such states, as it were. Knowledge states only correspond to partitions of the "external" features of the cube.

4.5. THE GENERAL CASE $L > 1$

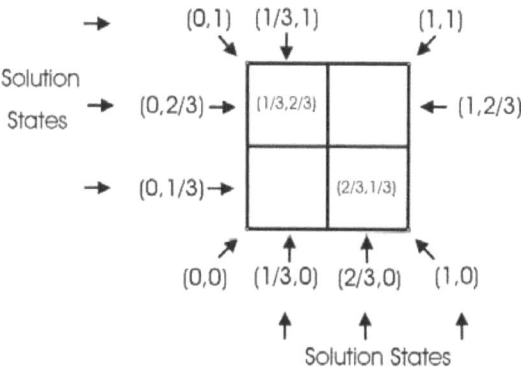

Figure 4.3: The case $L = 2, n = 2$

Game III: $L = 2, n = 3$

To further illustrate the general case, we give a sample game with two lies (4-valued logic) and a search space of 3 elements.

Game III.
Let the unknown $x = 2$
$K0 = (1, 1, 1)$
Question 1: $x \in \{1, 2\}$?
Answer: no (first lie)
$K1 = (2/3, 2/3, 1)$
Question 2: $x \in \{1, 3\}$?
Answer: no
$K2 = (1/3, 2/3, 2/3)$
Question 3: $x \in \{3\}$?
Answer: yes (second lie)
$K3 = (0, 1/3, 2/3)$ (note that at this stage, the answer $x = 1$ can already be ruled out)
Question 4: $x \in \{2\}$?
Answer: yes
$K4 = (0, 1/3, 1/3)$
Question 5: $x \in \{3\}$?
Answer: no
$K5 = (0, 1/3, 0)$

This is a solution state so the solution $x = 2$ is read off. The non-zero component even shows how many lies were told. Had it been 1 then no lies would have been told, and if it had been 2/3 then one lie would have been told. See Figure 4.4.

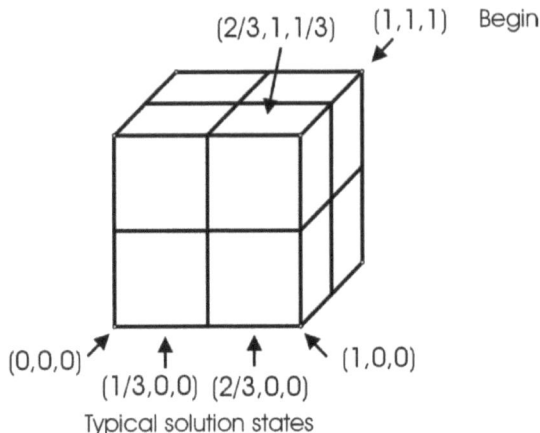

Figure 4.4: Game III: $L = 2, n = 3$

For general L and n, the n–cubes can be broken up into arrays of L^n sub-n-cubes, by separating along the lines drawn on the faces. Fig. 4.5 shows the case $L = 3, n = 3$.

To avoid being misleading, recall from above that not every edge and face is a possible knowledge state, since the partitions drawn on the 2-faces, as well as the internal faces and cubes, are not such. In passing, this suggests a paraconsistent extension of Ulam games where more than one Łukasiewicz value is assigned to the same component in a knowledge state, which could be achieved by allowing multiple inconsistent answers to questions, say by allowing more than one answerer, and assigning answers to boundaries between consistent faces. This possible development is not pursued here.

4.6 Paraconsistency

There is an obvious intuitive sense in which Ulam games have the flavour of paraconsistency about them. But what is it? After all, Łukasiewicz logics are not paraconsistent, or at least they are not paraconsistent if the top value $\{1\}$ is the sole designated value, as we have already noted. They can be made paraconsistent by extending the designated

4.6. PARACONSISTENCY

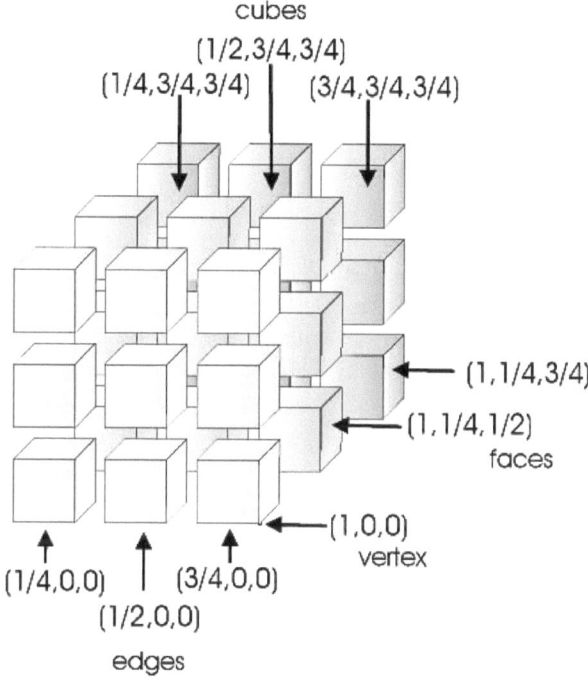

Figure 4.5: The case $L = 3, n = 3$

values to include a negation fixed point, but that rather trivialises the point. Mundici (2002, 407) points out that Ulam games are fault-tolerant (paraconsistent) in the sense that the procedure allows successive opposite answers to the same question to yield a knowledge state which is not impossible. If we define the *Pinocchio negation* $\neg$ of a knowledge state as that knowledge state brought about by an opposite answer to the same question, then we can see that $x \odot \neg x$ need not be an impossible knowledge state.

This is a good point, but it is also a little bit misleading. First, we should distinguish *atomic* knowledge states, as vectors of values which are either $L/(L+1)$ or 1. These are the knowledge states produced by answers to a single question, before fusing with the pre-existing knowledge state. Now there is a natural Pinocchio negation definable on the atomic knowledge states: the vector which interchanges the 1s with the $L/(L+1)$s. This corresponds to the opposite answer to the same question. Pinocchio negation is not to be confused with Łukasiewicz negation, where the components are subtracted from 1.

Unfortunately, Pinocchio negation resists extension to general knowledge states, at least if reasonable properties for negation, such as Double Negation, are to be preserved. It appears to be a plausible negation operator only on atomic knowledge states. If so, then the observation that $x \odot \neg x$ is not an impossible knowledge state only makes sense when x is confined to atomic states.

But now it is apparent why this should be so. Atomic knowledge states correspond 1-1 to the characteristic functions of subsets of the search space S. For example, let $S = \{1, ..., 5\}$, and ask whether $x \in \{1, 2, 3\}$? The answer *yes* has the atomic knowledge state $(1, 1, 1, L/(L+1), L/(L+1))$, while the answer *no* has the atomic knowledge state $(L/(L+1), L/(L+1), L/(L+1), 1, 1)$, which is its Pinocchio negation. Thus, *Pinocchio negation is but classical negation on the set of atomic knowledge states,* and these themselves form the usual Boolean set algebra under set unions and intersections.

This does not detract from the fault-tolerant character of the Ulam game operators. But interestingly, it re-locates it, in the *interaction* between an essentially Boolean negation $\neg$ and the Łukasiewicz fusion $\odot$. It makes the paraconsistency more a matter of the properties of conjunction, an approach which has certainly recommended itself to various significant scholars of paraconsistency. It can be speculated as to how widespread this phenomenon might be.

In conclusion, while we have presented the geometrical representations as arising out of Ulam games, one can view the situation conversely, as presenting Łukasiewicz logics as arising out of geometrical figures. Then the Ulam games serve to supply the linking functor for these connections.

*

In the next 3 chapters, we return to the theme of groups. In Chapter 5 we see how inconsistent theories arise from groups of symmetries. In Chapter 6 a general result is shown which relates homomorphisms to inconsistent theories. Then in Chapter 7 this is applied to a major branch of geometry, namely algebraic topology.

Chapter 5

Symmetry

5.1 The Idea of Symmetry

An *isometry* is a distance-preserving operation, such as a *rotation, reflection* or *translation*. In this chapter we will consider isometries in the (2-D) Euclidean plane. An isometry of a body, or a figure, or a space, is called a *symmetry* if it leaves the figure, or the space, in some sense unchanged. This is expressed informally in various ways. For example, Birkhoff and Mac Lane speak of the square as having rotational symmetry, saying vaguely that there are rotations in which the square is "carried into itself" (1965, 110). Also Coxeter refers to isometries as symmetries when they "leave the whole figure unchanged while permuting its parts" (1961, 30). This is at least misleading, and arguably false: the symmetries of the square are strictly a subset of the figure with permuted vertices: there are permutations of the vertices of the square which carry the square to itself but which are not isometries.

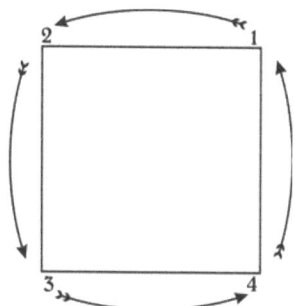

Figure 5.1: Rotational symmetries of the square

This makes for an interesting matter as to how to make these informal metaphors precise and accurate. After all, any figure whatsoever can trivially be carried into itself isometrically; that is by being rotated and then have its parts associated 1-1 with the original. (Those of you who are confident about the correct statement here are invited to submit.) But at any rate, many figures are intuitively not symmetrical. Hence it is necessary to describe those that are.

The idea of a rotational symmetry involves that of sameness and difference, on two levels. At the first level, there is the general idea of a rotation. Here, a single body, or figure, rotates, the **same** body or figure before and afterwards, even though the space it occupies changes. In contrast, rotations **differ** from one another. Rotations are transformations. They are indexed by angles θ, where $0 \leq \theta < 2\pi$ (radians). Different angles imply different rotations, so there are nondenumerably many distinct rotations. Rotations have an additive arithmetic. It is that of the (infinite) additive group of real numbers modulo 2π.

At the second level, there is the notion of a symmetry. The body or figure occupies the **same** region, and there may be **different** rotations in which the body or figure occupies the same region. For example, the square has just four nonidentical rotational symmetries, $0 \neq \pi/2 \neq \pi \neq 3\pi/2$. Symmetries have their own additive arithmetic. For the square, it is isomorphic with the (finite) additive group of integers modulo 4, that is Z_4. This is the *symmetry group of rotations* of the square.

In sum, one supposes a body or figure occupying a set S of points of a space, and after a rotation θ the same body or figure occupies a possibly different set S' of points. Then θ is a symmetry of the body iff $S = S'$. It is envisaged that the same parts of the body or figure might occupy different parts of space before and after the transformation, however for symmetry the same parts of space are occupied by some part or other of the body or figure.

5.2 An Inconsistent Approach to Rotational Symmetries

One way to describe the structure of the rotational symmetries of the square is to say that it is isomorphic with the finite additive group of integers Z_4. This is a cyclic Abelian group of order 4, having a single generator. Often enough, the language of multiplicative groups is used to describe these four transformations, but addition is a better analogy for angles so we use that. It is well known that the $(+, -)$ arithmetic for Z_4 can be described in a consistent theory. We take a different approach here, to illustrate the above remarks about sameness and difference by means of an inconsistent theory. Nothing is lost, however, since something isomorphic with consistent Z_4 can be recovered. Indeed, this will be done by employing the Routley star operation, which is an important tool in the semantics of paraconsistent logics, thereby demonstrating a further connection between the concepts of paraconsistency and geometrical concepts, in this case symmetry.

So, we will commence with the basic consistent theory distinguishing rotations one from the other. We call this *the consistent theory of rotations*. This is real number additive arithmetic modulo 2π. In order to tolerate contradictions, 3-valued closed set logic is used as background logic (see Chapter 1, Section 2). Using this, we will extend the theory by adding what from the point of view of rotations are falsehoods, but which reflect the samenesses inherent in the idea of symmetries, namely $0 = \pi/2 = \pi = 3\pi/2$. Since the theory we began with also contains $0 \neq \pi/2 \neq \pi \neq 3\pi/2$, the new theory is inconsistent. It is easily seen to be non-trivial. It is also demonstrably functional, having the additive group structure of the reals modulo $\pi/2$.

5.3 A Formal Theory

To formalise these consistent and inconsistent theories of the rotations of the square, we begin with a language $\mathcal{L}$, consisting of: (a) *atomic terms*, or *names*, for every rotation θ, where $0 \leq \theta < 2\pi$. These can be taken as naming themselves. (b) *term-forming operators*, the additive group operations $\{+, -\}$. (c) *atomic sentences*, if t_1, t_2 are terms, then $t_1 = t_2$ is an atomic sentence. (d) *sentence operators* $\{\&, \vee, \neg\}$, with material implication $\supset$ and material equivalence $\equiv$ defined in the usual way. (e) *quantifiers* $\exists, \forall$.

For background logic we take 3-valued CSL from Chapter 1, renamed $P3$ which is its usual name. It is also convenient to rename the values as $\{T, B, F\}$, with the first two being designated values. The natural lattice order on these values is subset inclusion, which gives $F < B < T$.

For convenience in this chapter, we use Greek ϕ, ψ for metalinguistic propositional variables. We then define an interpretation I: (sentences of $\mathcal{L}$) $\rightarrow \{T, B, F\}$ as follows:

(i) if $(t_1 = t_2)$ mod 2π then $I(t_1 = t_2) = T$

(ii) else, if $(t_1 \neq t_2)$ mod 2π but $(t_1 = t_2)$ mod $\pi/2$, then $I(t_1 = t_2) = B$

(iii) else, $I(t_1 = t_2) = F$

(iv) $I(\phi\&\psi) = \min(I(\phi), I(\psi))$

(v) $I(\phi \vee \psi) = \max(I(\phi), I(\psi))$

(vi) $I(\neg T) = F$, $I(\neg B) = I(\neg F) = T$

(vii) $I((\forall x)Fx) = \min\{v : \text{for some term } t, I(Ft) = v\}$

(viii) $I((\exists x)Fx) = \max\{v : \text{for some term } t, I(Ft) = v\}$

Definition 29 *The inconsistent theory of rotational symmetries of the square, Th_{RS}, is defined to be $\{\phi : I(\phi) \in \{T, B\}\}$.*

We note:

(a) The "background theory", the consistent theory of rotations, is given by $\{\phi : I(\phi) = T\}$

(b) In virtue of clauses (ii) and (vi), Th_{RS} is inconsistent, since it contains all of $0 = \pi/2$, $\neg(0 = \pi/2)$, $0 = \pi$, $\neg(0 = \pi)$, $0 = 3\pi/2$, and $\neg(0 = 3\pi/2)$.

(c) In virtue of clause (i), all true identities hold consistently, including $0 = 0$ and $\pi/2 = \pi/2$.

(d) From (c), $\neg 0 = 0$ does not hold, nor does $0 = \pi/3$ hold; so Th_{RS} is non-trivial.

Now, an important property for a theory to have is *functionality*:

Definition 30 *A theory is functional iff, whevever $t_1 = t_2$ holds, then Ft_1 holds iff Ft_2 holds, where Ft_1 and Ft_2 are atomic sentences, and Ft_2 is like Ft_1 except for substituting t_2 for t_1 in one or more places. A theory is transparent iff the same thing holds, but Ft_1 can be any context, not only atomic.*

Transparency implies functionality but not conversely, so transparency is a stronger concept. However, it is functionality, that is substitutivity of identicals in all atomic contexts, which is more important for a mathematical theory, for it is essential to be able to do calculations in a theory. Now we have:

Theorem 31 *Th_{RS} is functional.*

5.3. A FORMAL THEORY

Proof. We show this for a single occurrence of t_1 replaced by t_2. If Ft_1 is atomic, it has the form of a string of positive and negative names joined by $+$, equated with another such string, and containing one occurence of t_1. Addition is associative and commutative, so if Ft_1 holds in the theory, this can be reduced to one of the equations $t+t_1 = 0$ or $t-t_1 = 0$ holding in the theory. If an equation holds, then by the construction of I, it holds either in mod 2π or in mod $\pi/2$. The former implies the latter, so any equation holding in this theory holds in mod $\pi/2$. Hence, if in addition $t_1 = t_2$ holds, it holds in mod $\pi/2$, as does $-l_1 = -l_2$. But additive arithmetic mod $\pi/2$ permits substitutivity of identicals, so $t + t_2 = 0$ or $t - t_2 = 0$ respectively hold in mod $\pi/2$. The construction of Th_{RS} then ensures that this equation holds in the theory, and by reassembling terms Ft_2 holds.

For more than one occurrence of t_1 being replaced, the result follows by the obvious mathematical induction. ∎

In contrast with the previous result, we note that Th_{RS} is not transparent: both $0 = \pi/2$ and $\neg(0 = \pi/2)$ hold, so that transparency would allow that the former could be substituted for in the latter, ensuring that $\neg(0 = 0)$ holds, which it does not. As noted above, this failure of transparency does not seem to be too problematic. Indeed, there are ways to make theories transparent, particularly by using a different background paraconsistent logic such as $RM3$ instead of $P3$, but we do not take that up here.

Finally, we come to the question of *extendability*. One theory is an *extension* of another iff the latter is a subset of the former. Various extendability results are known, e.g. for $RM3$ (see e.g. Mortensen 1995, 23-4) and Priest's LP. However, proofs of extendability are sensitive to background logic. In particular, the choice of $P3$ as background logic requires careful construction of the interpretation function I. In the present case, extendability is demonstrable, but we need a lemma.

Lemma 32 *(Extendability Lemma for P3).* Let Th_1, Th_2 be theories in the same language generated by interpretations I_1, I_2 into the topological logic P3. If (a) the set of atomic sentences holding consistently in $Th_1 \subseteq$ the set of atomic sentences holding consistently in Th_2, and (b) the set of atomic sentences holding inconsistently in $Th_1 \subseteq$ the set of atomic sentences holding inconsistently in Th_2, and (c) the set of atomic sentences failing in $Th_1 \subseteq$ the set of atomic sentences either failing or holding inconsistently in Th_2, then $Th_1 \subseteq Th_2$, that is Th_2 is an extension of Th_1.

Proof. By induction on the number of occurrences of $\{\&, \vee, \neg, \exists, \forall\}$. Note that the hypothesis of the theorem is equivalent to: For any atomic

ϕ, (a) $I_1(\phi) = T$ implies $I_2(\phi) = T$, (b) $I_1(\phi) = B$ implies $I_2(\phi) = B$ and (c) $I_1(\phi) = F$ implies $I_2(\phi) \in \{B, F\}$. The induction proves it for all sentences. The base clause is the hypothesis of the theorem. Inductive clauses:

($\neg$ case:) (a) If $I_1(\neg\phi) = T$ then $I_1(\phi) \in \{B, F\}$, so by IH $I_2(\phi) \in \{B, F\}$, so $I_2(\neg\phi) = T$. (b) $I_1(\neg\phi) = B$ is impossible by the table for $\neg$. (c) If $I_1(\neg\phi) = F$ then $I_1(\phi) = T$, so by IH $I_2(\phi) = T$, so $I_2(\neg\phi) = F \in \{B, F\}$.

(& case:) (a) If $I_1(\phi\&\psi) = T$ then $I_1(\phi) = I_1(\psi) = T$, so by IH $I_2(\phi) = I_2(\psi) = T$, so $I_2(\phi\&\psi) = T$. (b) If $I_1(\phi\&\psi) = B$ then one of $I_1(\phi), I_1(\psi) = B$ and the other $\in \{T, B\}$. By IH, one of $I_2(\phi), I_2(\psi) = B$ and the other $\in \{T, B\}$, so $I_2(\phi\&\psi) = B$. (c) If $I_1(\phi\&\psi) = F$ then one of $I_1(\phi), I_1(\psi)$ is F, so by IH one of $I_2(\phi), I_2(\psi) \in \{B, F\}$, so $I_2(\phi\&\psi) \in \{B, F\}$.

($\vee$ case:) (a) If $I_1(\phi \vee \psi) = T$ then one of $I_1(\phi), I_1(\psi) = T$, so by IH one of $I_2(\phi), I_2(\psi) = T$, so $I_2(\phi \vee \psi) = T$. (b) If $I_1(\phi \vee \psi) = B$ then one of $I_1(\phi), I_1(\psi) = B$ and the other $\in \{B, F\}$, so by IH one of $I_2(\phi), I_2(\psi) = B$ and the other $\in \{B, F\}$, so $I_2(\phi \vee \psi) = B$. (c) If $I_1(\phi \vee \psi) = F$ then $I_1(\phi) = I_1(\psi) = F$, so by IH $I_2(\phi) = I_2(\psi) = F$, so $I_2(\phi\&\psi) = F \in \{B, F\}$.

($\forall, \exists$ cases:) These are similar to the & and $\vee$ cases respectively. ∎

From the lemma, it immediately follows that:

Theorem 33 *Th_{RS} is an extension of the consistent theory of rotations.*

Proof. The construction of the interpretation for Th_{RS}, in comparison with the consistent theory of rotations, exactly satisfies the conditions of the lemma. ∎

5.4 Recovering the Consistent Theory of Rotational Symmetries

To justify where we have been, we show how to get back to consistent symmetries. We use the Routley Functor. As we noted before, the Routley Functor, also known as the Routley Star Operator, is an important device in the semantics of relevant and other paraconsistent logics. This bears on our main theme, that there are significant interactions here between logical and geometrical concepts.

The *consistent theory of the rotational symmetry group* has four distinct symmetries $0 \neq \pi/2 \neq \pi \neq 3\pi/2$. These four symmetries corre-

5.4. RECOVERING THE CONSISTENT THEORY ...

spond to four identities $0 = 0$, $\pi/2 = \pi/2$, $\pi = \pi$, $3\pi/2 = 3\pi/2$. So we want to extract from the inconsistent theory a cut-down consistent theory that has these identities and no others, together with their group operations. This is done in two steps. First, we eliminate all use of terms other than the terms built up from the four symmetries. To do this, form a cut-down language which we can call $\mathcal{L}_4$, in a like fashion to the formal language constructed at the beginning of Section 3 of this chapter, save that in the first clause we have just the four names $\{0, \pi/2, \pi, 3\pi/2\}$, and terms built from them with $\{+, -\}$. Intersecting this with Th_{RS} gives the theory of rotational symmetries purged of reference to rotations other than just these four.

This theory is, however, inconsistent, since it contains, for example, both $0 = \pi/2$ and $\neg 0 = \pi/2$. Now we can use the Routley functor.

Definition 34 *Let S be any set of sentences, then S^*, the Routley star of S, is defined as $S^* = \{\phi : \neg\phi \notin S\}$.*

The effect of the star operator on inconsistent theories is to snip out sentences whose negation is also in the theory. When Th is a theory of group logics or De-Morgan style logics such as $RM3$, in which Double Negation DN holds (A is interdeducible with $\neg\neg A$), then it contains arbitrary iterations of negations of any sentence holding inconsistently. So the effect of star here is to remove all inconsistent pairs $\{\phi, \neg\phi\}$, rendering the theory incomplete. When Th is a $P3$-theory, on the other hand, the only possibilities for inconsistent pairs are generated by sentences ϕ taking the middle value B. However while the negations of these $\neg\phi$ take the top value T, the next iteration of negation $\neg\neg\phi$ takes the value F and so is not in Th. Hence, starring $P3$-theories cuts out the ϕ of any inconsistent pair, but leaves $\neg\phi$, $\neg\neg\neg\phi$, etc. consistently in Th^*. For more details, see (*Inconsistent Mathematics*, Chap 13). This gives us:

Theorem 35 *The consistent theory of rotational symmetries of the square is given by $(Th_{RS} \cap \mathcal{L}_4)^*$.*

Proof. It is clear that this theory is consistent, and that in it we have $0 = 0$, $\pi/2 = \pi/2$, $\pi = \pi$, $3\pi/2 = 3\pi/2$, and no other identities with names; also we have the disidentifications $0 \neq \pi/2 \neq \pi \neq 3\pi/2$, as well as all the correct identities and disidentities involving group sums and differences. ■

5.5 Symmetries of Reflection

So far in this chapter we have concentrated on the rotational symmetries of the square. By implication, the rotational symmetries of any regular n-sided polygon can be treated in the same way, save that the angles of symmetry are $2\pi/n$ instead of $\pi/2$. But these are not the only symmetries of the square. There are also the symmetries of reflection. There are four of these for the square, corresponding to four axes of reflection, which suggest the names H, V, D, D',

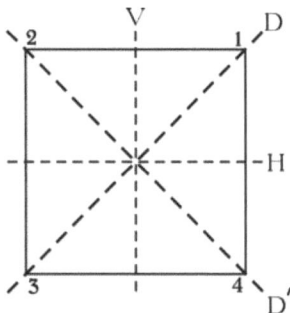

Figure 5.2: Reflective symmetries of the square

Each of these reflections can be treated in a similar fashion to the rotational symmetries, save that the reflections can be thought of as rotations through π in the third dimension. Since each of these reflections returns to the original when repeated, each has its own symmetry subgroup isomorphic with Z_2. Application of the reflection twice returns us to the identity 0, corresponding to $1 + 1 = 0$ in Z_2.

5.6 The Dihedral Group

Added to the four rotational symmetries, the four reflections give us eight symmetries in all, and that is the total number of symmetries for the square. This group of order eight is known as the *dihedral group* D_4. It can be seen that the eight symmetries of the square form a group by applying them in pairs and checking that the outcome is always an element of the group, and that for any operation there is an operation which can serve as its negative since it returns to the identity 0. It is noted that this group is not Abelian, for example rotation through $\pi/2$ followed by reflection in the horizontal axis gives a different result if the order of the operations is reversed. In general, the symmetries of the regular $2n$-sided polygon form the dihedral group D_n of order $2n$.

5.6. THE DIHEDRAL GROUP

There is a useful way to represent the dihedral group, employing the idea of a direct product of its subgroups.

Definition 36 *For any two groups G, H, the direct product $G \times H$ is defined to be the group whose elements are the ordered pairs (g, h) where $g \in G$, $h \in H$, with the group operation definable pointwise, that is $(g, h) + (g', h') = (g + g', h + h')$. The identity is $(0, 0)$, and $-(g, h) = (-g, -h)$.*

When G is Z_4 and H is Z_2, $G \times H$ is order 8. The pairs can be thought of as corresponding to two operations, where first element of each pair gives the rotation, followed by the second operation which is a reflection. Any one of the four reflections can fill this role, and the effect of applying a reflection is to reorder the vertices including reversing the sense of the vertices (clockwise to anticlockwise). Any one of the four reflections, in combination with the four rotations, suffices to generate all 8 symmetries.

It is clear that there are subgroups $\{(g, 0) : g \in G\}$ isomorphic with G, and $\{(0, h) : h \in H\}$ isomorphic with H. There are also two *projection homomorphisms* $P_G : G \times H \to G$ with $P_G((g, h)) = g$, and $P_H : G \times H \to H$ with $P_H((g, h)) = h$. Now it is well-known that a homomorphism generates congruence classes of elements sent to the same place by the homomorphism. So for example, P_H with h fixed identifies the (distinct) four elements of the rotational group of symmetries. Sending these to the middle value B ensures that their disidentities hold as well. This leaves two equivalence classes, one for $P_H = 0$, and one for $P_H = 1$. The resultant theories in the language $\mathcal{L}_4$ are inconsistent, nontrivial and functional. There is also a corresponding result for the other projection function P_G.

*

There is another type of symmetry in the plane, translational symmetry. This is exemplified by infinite figures forming a repeating tesselation of the plane. We do not take this up here, but it is clear that (1-D) translations form a group of symmetries, and that manouvres similar to those above can be made to generate nontrivial inconsistent functional theories.

In the next chapter, we look more closely into the role of homomorphisms in generating inconsistent functional theories. In turn, this will clear the way for a discussion of homological algebra in the following chapter.

Figure 5.3: Rosette

Chapter 6

Homomorphisms

6.1 General Algebras

At the end of the last chapter we came across a link between homomorphisms and inconsistent theories. In this chapter the idea is developed further. This is a preliminary to considering the role of group homomorphisms in homological algebra, which we turn to in the next chapter. In this chapter, we consider first general algebras, then in the second section apply the results to groups.

Recalling basic concepts: an *algebra* is a set equipped with a collection of n-ary operations (*ie.* functions) from the set to itself, but not necessarily onto. An algebra $\mathcal{A}$ on a set A can be written $\mathcal{A} = \{A, \{\circ_\varphi\}_{\varphi \in \Phi}\}$, where the set Φ indexes the operations $\circ_\varphi$. An important special case is where Φ is finite, when one writes $\{A, \circ_1, ..., \circ_n\}$. Algebras are sometimes referred to loosely by their corresponding sets. Algebras may be described by equational theories at the atomic level, closed under the usual logical operators as in earlier chapters. A subset $\mathcal{A}'$ of $\mathcal{A}$ is a *subalgebra* of $\mathcal{A}$, if $\mathcal{A}'$ is closed under all the operations of $\mathcal{A}$. The intersection of any class of subalgebras of $\mathcal{A}$ is also a subalgebra of $\mathcal{A}$. A subset $\mathcal{A}' \subseteq \mathcal{A}$ is a *set of generators* for $\mathcal{A}$, if the whole algebra $\mathcal{A}$ is the only subalgebra containing $\mathcal{A}'$.

Two algebras are *similar* if their operations are indexed by the same set, and corresponding operations have the same arity. The corresponding operations of similar algebras are often referred to by the same symbols. If $\mathcal{A}, \mathcal{B}$ are similar algebras, then a *homomorphism* from $\mathcal{A}$ to $\mathcal{B}$ is a generally many-one, into or onto, function $h : \mathcal{A} \to \mathcal{B}$ which preserves all the operations; that is $h(\circ_\varphi(a_1, ..., a_n)) = \circ_\varphi(h(a_1), ..., h(a_n))$. If h is also 1-1 and onto it is an *isomorphism*.

Now we can prove a result essentially due to Dunn (1979), see also Priest (2006, 227).

Theorem 37 *Let $h : \mathcal{A} \to \mathcal{A}'$ be a homomorphism, where $\mathcal{A}'$ is a subalgebra of $\mathcal{A}$. Then the consistent equational theory of $\mathcal{A}$ has an inconsistent functional extension.*

Proof. We take as atomic names all the members of $\mathcal{A}$, and for nonatomic terms close the atomic terms under all the operations of $\mathcal{A}$. Then make the following interpretation I in the 3-valued logic $P3$:

(i) $I(t_1 = t_2) = T$ iff $t_1 = t_2$ in $\mathcal{A}$

(ii) $I(t_1 = t_2) = B$ iff $t_1 \neq t_2$ in $\mathcal{A}$ but $h(t_1) = h(t_2)$

(iii) Otherwise $I(t_1 = t_2) = F$.

The clauses for the logical operators are as in $P3$.

This defines a theory Th_h by $Th_h = \{\phi : I(\phi) \in \{T, B\}\}$. We note that the consistent equational theory of $\mathcal{A}$ is given by dropping clause (ii). For convenience we name that $Th_\mathcal{A}$.

First we have that Th_h is an extension of $Th_\mathcal{A}$. This is an immediate consequence of the fact that the definition of I satisfies the conditions of the extendability lemma for $P3$ (previous chapter).

Second, Th_h is inconsistent when h is many-one, since then h identifies elements that are distinct back in $\mathcal{A}$, and so I sends their equation to the middle value B.

Third, Th_h is functional, that is preserves substitutivity of atomic equations in all atomic contexts. This can be shown by an inductive argument on the numbers of substitutions of one term for another in atomic contexts. We give the base clause, the inductive step is similar. If an equation holds, it has the value T or B. If $t_1 = t_2$ has the value T, since any operator $\circ$ is a function, $\circ(t_1, a_2, ..., a_n) = \circ(t_2, a_2, ..., a_n)$, so that this equation is in Th_h. Otherwise, if $t_1 = t_2$ takes the value B, then $h(t_1) = h(t_2)$. Hence, since $\circ$ is a function, $\circ(h(t_1), h(a_2), ..., h(a_n)) = \circ(h(t_2), h(a_2), ..., h(a_n))$. But h is a homomorphism, so $h(\circ(t_1, a_2, ...a_n)) = h(\circ(t_2, a_2, ...a_n))$. By the construction of I, then, the equation $\circ(t_1, a_2, ..., a_n) = \circ(t_2, a_2, ..., a_n)$ takes the value B, and so is in Th_h. ∎

We observe that Th_h is nontrivial, since for example negations of equations getting the top value T will fail to be in the theory. However, the theory can be *mathematically trivial* (all atomic sentences holding), when for example h identifies all elements. But it will not be mathematically trivial if h does not identify all elements. As with earlier constructions, Th_h is not transparent. However, a different construction is transparent: use any paraconsistent logic where $\neg B = B$, such as $RM3$ or LP, and send all equations that hold to B. Finally, applying the Routley functor to Th_h recovers $Th_{\mathcal{A}}$, in a like fashion to the previous chapter. (Exercises: fill in the details of the above.)

It is well known that the subalgebra $\mathcal{A}'$ is isomorphic with the collection of congruence classes $[\mathcal{A}] = \{[a] : a \in \mathcal{A}\}$. The difference between Th_h and $Th_{\mathcal{A}'}$ is that Th_h registers which of the identities in $\mathcal{A}'$ are lifted up from being disidentical in $\mathcal{A}$, by combining such identities and disidenties, that is by having the identities hold inconsistently. While $\mathcal{A}'$ is generally smaller than $\mathcal{A}$, Th_h is generally larger than $Th_{\mathcal{A}}$, the consistent theory of $\mathcal{A}$. If we call the set of equations holding in $\mathcal{A}'$ the *algebra* of $\mathcal{A}'$, we can also say, in virtue of these equations all taking T or B in I, that Th_h is also an extension of the algebra of $\mathcal{A}'$. One should not, however, go on to conclude that Th_h is an extension of the full (nonatomic) theory of $\mathcal{A}'$: the Extendability Lemma breaks down because $\mathcal{A}$ and $\mathcal{A}'$ do not share the same language, the latter language has generally fewer names than the former, which makes for a difference in which quantificational sentences hold.

Again, if $h : \mathcal{A} \to \mathcal{B}$ is any homomorphism, then h is the composition of a homomorphism $i : \mathcal{A} \to \mathcal{A}'$ and an isomorphism $j : \mathcal{A}' \to \mathcal{B}$, that is $h(x) = j(i(x))$. The isomorphism produces isomorphic equational theories of $\mathcal{A}'$ and $\mathcal{B}$, while the homomorphism picks up the inconsistent identities of the congruence classes to produce Th_h.

6.2 Groups

All of the above results continue to obtain when the homomorphism in question is a group homomorphism, but there is more to say because of the arithmetic structure that groups possess. Revising some basic definitions:

If $\mathcal{G}'$ is a subgroup of $\mathcal{G}$, then a *right coset* of $\mathcal{G}'$ is any set $\{\mathcal{G}' + a\}$ of right hand sums of all the members of $\mathcal{G}'$ with a fixed member a of $\mathcal{G}$; and a *left coset* is any set $\{a + \mathcal{G}'\}$ of left hand sums etc.

A subgroup is *normal* iff all its right cosets are left cosets and vice versa, in which case we drop the words "right" and "left". It is clear that every subgroup of an Abelian group is normal. Two cosets of the same subgroup are either identical or non-overlapping.

If $h : \mathcal{G} \to H$ is a group homomorphism, then the set $\{a : h(a) = 0\}$, that is the set of elements taken to zero by h, is a normal subgroup $\mathcal{K}$ called the *kernel* of h. That is, the kernel $\mathcal{K}$ is the congruence class of zero, [0].

Consider the cosets of the kernel of a homomorphism. Two elements of $\mathcal{G}$ have the same image under h iff they are in the same coset of the kernel. To take an example, consider the homomorphism from Z to Z_4. The kernel $\mathcal{K} = [0] = \{\ldots -8, -4, 0, 4, 8, \ldots\}$ is one of the cosets. The other cosets of $\mathcal{K}$ are the three congruence classes $\{\ldots -7, -3, 1, 5, 9, \ldots\}$, $\{\ldots -6, -2, 2, 6, 10, \ldots\}$ and $\{\ldots -5, -1, 3, 7, 11, \ldots\}$. For example, $h(-7) = h(-3) = h(1) = h(5) = h(9) = \ldots$etc., and similarly for the other cosets.

More generally, cosets can be constructed for any normal subgroup $\mathcal{N}$ of a group, independently of the presence of a homomorphism. We can define the relation $a \approx b$ to mean $a - b \in \mathcal{N}$. (In the notation of mutiplicative groups this is written $ab^{-1} \in \mathcal{N}$). It is easy to see that $\approx$ is a congruence on the group operations, and that if h is a homomorphism of which $\mathcal{N}$ is the kernel, then $a \approx b$ iff $h(a) = h(b)$.

Cosets admit of group operations. The sum of the cosets $\{\mathcal{N} + a\}$ and $\{\mathcal{N} + b\}$ can be defined as $\{\mathcal{N} + a + b\}$, and the identity as $\{\mathcal{N} + 0\}$. Hence we can make a definition:

Definition 38 *If $\mathcal{N}$ is any normal subgroup of $\mathcal{G}$, then the group of cosets of $\mathcal{N}$ is called the **quotient** group, or **factor** group, or **difference** group, of $\mathcal{G}$ by $\mathcal{N}$. In additive notation it is written $\mathcal{G} - \mathcal{N}$, and in multiplicative notation it is written $\mathcal{G}/\mathcal{N}$.*

That is, the additional arithmetical machinery of groups allows an algebra of congruence classes to be constructed without appeal to a homomorphism. However, note that, having constructed the factor group from a normal subgroup $\mathcal{N}$, a homomorphism $h : \mathcal{G} \to \mathcal{G} - \mathcal{N}$ is recoverable by $h(a) := \{\mathcal{N} + a\}$. Thus the homomorphic images of a given group are exactly the factor groups by its different normal subgroups.

In terms of the above example: the multiples of 4 form a normal subgroup N_4 of Z, $N_4 = \{\ldots -8, -4, 0, 4, 8, \ldots\}$. The other cosets are therefore also the same as above: $\{\ldots -7, -3, 1, 5, 9, \ldots\}$, etc. The factor group operations defined on the cosets correspond exactly to those of Z_4, so that $Z - N_4$ is isomorphic with Z_4. But also, the multiples of any natural number n form a normal subgroup N_n of Z, and therefore $Z - N_n$ is isomorphic with Z_n. In each case, the various homomorphisms $h : Z \to Z - N_n$ are given by sending a to the set containing a summed successively with the members of the corresponding kernel.

Here we can readily see:

Theorem 39 *There exist inconsistent functional theories that extend equational theories of factor groups.*

6.2. GROUPS

Proof. This is essentially the same technique as that already used, save that instead of a homomorphism we use the normal subgroup $\mathcal{N}$. Set $I(t_1 = t_2) = T$ iff $t_1 = t_2$ in G, set $I(t_1 = t_2) = B$ iff $t_1 \neq t_2$ but $t_1 - t_2 \in \mathcal{N}$, and otherwise set $I(t_1 = t_2) = F$. Membership of the theory is determined by taking one of the two designated values. The proofs of extendability, nontriviality and functionality are the same as in the previous section. ■

Again, these inconsistent theories track the origins of elements in the original group from which the factor group is derived. Identities in the factor group that are not identities in the original group hold inconsistently in the theory.

*

The foregoing is a useful way to think of the relationships between theories determined by a homomorphism from one algebra to another, or by a factor group arising from a normal subgroup of a group $\mathcal{A}$. The inconsistent theory extends the consistent theory of the algebra $\mathcal{A}$, including a "record" of what the (smaller) factor group additionally identifies in the algebra of $\mathcal{A}'$, in the form of inconsistencies.

Factor groups are a ubiquitous construction in geometry and elsewhere that groups are used. We see them in action again in the next chapter.

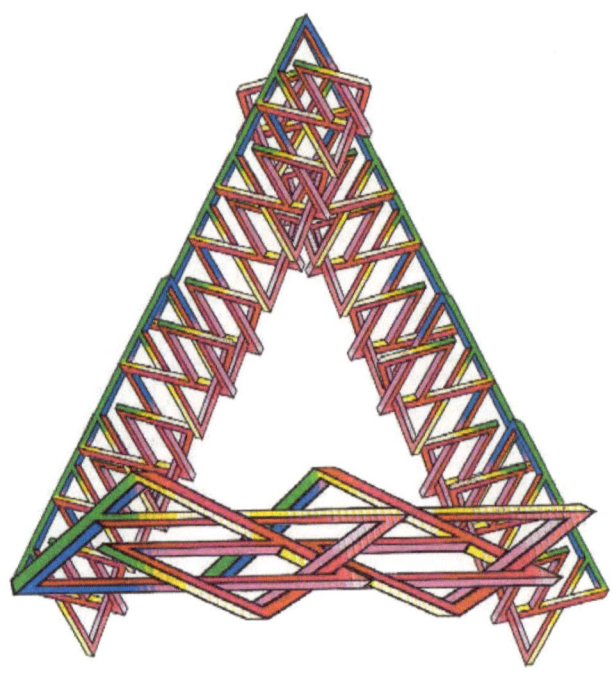

Figure 6.1: Triangular lattice

Chapter 7

Homology

7.1 Introduction

The reader will have noticed that techniques in the theory of inconsistency have an affinity with algebra; though this is not to deny the existence of inconsistent theories elsewhere in mathematics, of course. Hence, to find connections between inconsistency and geometry, one place to look is at algebras arising in geometrical theories. One salient area is *algebraic topology*. This seems to be mostly a twentieth-century development, indeed coming to be somewhat after general topology, which is what we looked at in earlier chapters. Algebraic topology has at least two rich specialities, homology theory and homotopy theory. In this chapter, first we survey homology theory for the benefit of logicians. This then is seen to be yet another place where inconsistency concepts interact with classical mathematics.

Homology theory aims to classify (n-dimensional) figures algebraically in such a way as to notice such things as how many holes the figure has (sphere = 0, torus =1 for example). Its distinctive flavour is that it addresses the problem in a rather computational way, by means of polygons and generally assemblages of figures with straight edges and flat (n-)faces. It proves possible to study these by means of (additive) groups. Moreover, since any figure can be triangulated arbitrarily closely, the theory clearly has more general application.

7.2 Simplices

The theory begins with a collection of basic or paradigm figures, called *simplices;* and then extends to assemblages of these, called *simplicial complexes*. The basic simplices are: (a) in zero dimensions, a point or vertex; (b) in one dimension, a line or edge; (c) in two dimensions, a triangle; (d) in three dimensions, a tetrahedron, and so on. Each of these is named by beginning with single symbols for the vertices, such as $v_0, v_1...$, then taking pairs of vertices such as $v_0 v_1$ for the lines, triples such as $v_0 v_1 v_2$ for the areas (triangles), quadruples such as $v_0 v_1 v_2 v_3$ for the volumes (tetrahedra), and so on. A complex can then be described by a group sum of its simple components. This begins to set up descriptions of arbitrary complexes in terms of groups. The (linear) order of the vertices imposes a natural *orientation*; for example, for a line, left-or-right; for a triangle clockwise-or-anticlockwise (see Figure 7.1). It turns out that this implies that the orientations of the edges of a triangle must not form a *cycle* (see below). Cycles do however play an important role in homology theory, as we see. When simplices are composed into a complex, it is necessary for edges being identified together to have the same orientation.

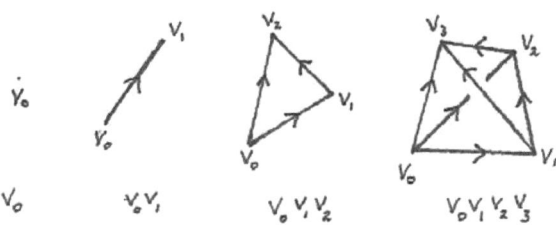

Figure 7.1: Point, line, triangle, tetrahedron; with orientations

Orientation enables the introduction of the idea of *group minus*. The group minus corresponds to the idea of adding an element in the opposite orientation. The group minus is clearly a kind of negation, as we have already seen in Chapter 3. Here it has a distinctive geometrical content: opposite directions are opposites of some kind. But this is not logic negation as it is standardly understood. When for example a line is summed with itself once as a positive and once as a negative, the result is the group zero, which represents "no net change". This is not logical contradiction. In the language suggested by connexive logic, it is "negation as cancellation".

7.2. SIMPLICES

The key concept in homology theory is the idea of the boundary of a figure, denoted by the boundary function ∂. As noted above, this is a development, made possible by groups, of the older idea of topological boundary. The boundary of a simplex is obtained by taking the complex (algebraic sum) of faces of the simplex one dimension down, made by deleting one vertex successively from the vertices of the simplex. There is an important proviso: that the boundary is obtained by an alternating sum of pluses and minuses of the faces taken in order. For example, given the tetrahedron $v_0v_1v_2v_3$, its boundary is $\partial(v_0v_1v_2v_3) - v_1v_2v_3$ $v_0v_2v_3 + v_0v_1v_3 - v_0v_1v_2$. Strictly, we have a sequence of boundary functions which are tracked by indexing the boundary function according to the dimensionality of the simplex, in the case of the tetrahedron it is ∂_3. Similarly for the triangle we write $\partial_2(v_0v_1v_2) = v_1v_2 - v_0v_2 + v_0v_1$. It is also natural to define higher-dimensional boundaries of lower-dimensional figures to be zero, that is $\partial_3(v_0v_1v_2) = 0$. Complexes then are dealt with term-by-term and then adding the results. Since the ∂_n so defined respect the group operations on the complexes, they are group homomorphisms.

In consequence of this construction, we have a central result of homology theory:

Theorem 40 $\partial_n\partial_{n+1} = 0$.

Proof. This means: first ∂_{n+1} then ∂_n. As an example, consider the result of applying first ∂_2 then ∂_1 to the group representation of the triangle. From above, $\partial_2(v_0v_1v_2) = v_1v_2 - v_0v_2 + v_0v_1$. But then $\partial_1(v_1v_2 - v_0v_2 + v_0v_1) = (v_2 - v_1) - (v_2 - v_0) + (v_1 - v_0) = 0$. The theorem is established by a general argument of a similar kind. ∎

This result is frequently expressed as "the boundary of a boundary is zero", and loosely written $\partial^2 = 0$. The zero is however the group zero, and should not be confused with the null set which is appropriate for topological boundaries. The group boundary of a simplex coincides with its topological boundary; however, reapplying boundary operations sees the outcomes come apart. For instance, in the three element closed set topological space $\{R, \{a\}, \{\}\}$, the topological boundary $\{a\}$ is its own boundary.

The machinery of group theory makes it possible to identify boundaries as *cycles*. Cycles are defined as themselves having zero boundaries, that is they are like "closed loops" (in various dimensions). This is intuitively plausible as an account of the surfaces of simplices and by extension complexes, and gives increased structure in comparison with topological boundaries. But now the concept of cycles makes possible the concept of *homological groups*. These give a way of comparing the n-dimensional components of a complex, for varying n, by cancelling out the effect of boundaries in dimensions higher than n.

To see this, consider a complex X, and identify all the n-dimensional components of X for the highest dimension n represented in the parts of X. Form the free group of these components, call it $C_n(X)$, the group of n-dimensional chains of X. Suppressing the X, the boundary operator $\partial_n : C_n \to C_{n-1}$ takes us to the free group of $(n-1)$-dimensional chains, which is in turn acted on by the next boundary operator ∂_{n-1}. Thus we have a sequence of homomorphisms:

$$ 0 \longrightarrow \dots \xrightarrow{\partial_{n+1}} C_n \xrightarrow{\partial_n} C_{n-1} \xrightarrow{\partial_{n-1}} \dots \longrightarrow C_1 \xrightarrow{\partial_1} C_0 \xrightarrow{\partial_0} 0 $$

The left hand arrow from 0 is inserted to identify the zero of the leftmost group. The whole sequence is called a *chain complex*.

The image of each homomorphism ∂_n is written $Im(\partial_n)$; and the kernel of ∂_n, the subgroup sent to zero by ∂_n, is written $Ker(\partial_n)$. Note that $Im(\partial_{n+1})$ is in the same group as $Ker(\partial_n)$, namely C_n. Furthermore, it can be shown from $\partial^2 = 0$, that $Im(\partial_{n+1})$ is a (normal) subgroup of $Ker(\partial_n)$. This allows us to form the homology groups:

Definition 41 *The n-th homology group of X, $H_n(X) := Ker(\partial_n) \,/\, Im(\partial_{n+1})$.*

This definition has a natural meaning. The kernel of ∂_n are the n-dimensional parts of the figure that themselves have no boundary, that is the cycles. But we want to cancel out the effect of boundaries of parts of the figure from higher dimensions. Hence the factor or quotient construction, which identifies n-dimensional parts as being in the same coset iff they "differ by a boundary", that is $x \approx y$ iff $x - y \in Im(\partial_{n+1})$, and then x and y are said to be *homologous*. In the special case of a figure X where $Im(\partial_{n+1}) = Ker(\partial_n)$ for each n, the chain complex above is said to be an *exact sequence*.

The case $n = 2$, that is the exact sequence:

$$ 0 \longrightarrow C_2 \xrightarrow{\partial_2} C_1 \xrightarrow{\partial_1} C_0 \longrightarrow 0 $$

is called a *short exact sequence*.

The series of homology groups enables differences to be shown between figures of different dimensions with or without holes. The homology groups are copies of various sums of Z and Z_n. For a flavour for how these turn out, here are a few cases:

7.3. THE ROUTLEY FUNCTOR ON GROUPS

homology group	point	triangle	tetrahedron	torus
H_0	Z	Z	Z	Z
H_1	0	Z	0	$Z \oplus Z$
H_2	0	0	Z	Z
H_3	0	0	0	0

7.3 The Routley Functor on Groups

We have already seen in earlier chapters some of the uses of the Routley functor. But now we propose a new, extended definition for it, as an operator on groups, and indeed on their elements. We see that this yields a different way of describing factor algebras. In turn, that will enable us to make connections between the two concepts of negation in play, that arising from topological boundaries and that arising in homology theory.

We recall the usual definition of the Routley functor, as an operation on sets of sentences: if S is any set of sentences, then $S^* := \{A : \neg A \notin S\}$. We noted in 5.6 that, if S is a theory of a logic with Double Negation, then the function of the star operation is to remove contradictory pairs of sentences $\{A, \neg A\}$ from the theory rendering the result incomplete. It is also easily shown that it replaces missing pairs rendering the theory inconsistent. However, there are two points to note about this definition of the Routley functor. *First*, there is the implicit background language $\mathcal{L}$. If $\mathcal{L}(S)$, the language of S, is a proper subset of $\mathcal{L}$, then re-inserting missing pairs which belong to $\mathcal{L} - \mathcal{L}(S)$ will insert all the pairs from $\mathcal{L}$ which are no part of $\mathcal{L}(S)$. This is entirely appropriate sometimes, but not always. Hence the definition must be relativised to $\mathcal{L}$: that is $\{A : A \in \mathcal{L} \ \& \ \neg A \notin S\}$. The original definition can be recovered by taking $\mathcal{L} = \mathcal{L}(S)$, and when the original definition is used, this is assumed. *Second,* it is useful to take into account something which makes no difference to this definition, but which makes for a slightly neater generalisation: $S = S - \{\}$. Thus the revised full definition for the *Routley Star on sets of sentences* is:

$$S^*_\mathcal{L} = \{A : A \in \mathcal{L} \ \& \ \neg A \notin S - \{\}\}.$$

When we move to groups, the obvious thought arises as to whether there is a generalisation, and if so, does it have analogous properties. The answer is yes, but with interesting differences too, as we now show. The generalisations of the above are the natural ones.

The condition we want the generalisations to satisfy is:

Definition 42 *For any group G and subgroup N of G, the Routley star of N relative to G, is $N^*_G := \{x : x \in G \ \& \ -x \notin N - \{0\}\}$. When*

$G = N$, we have the (absolute) Routley star of N, $N^* := \{x : x \in N$ & $-x \notin N - \{0\}\}$.

We observe that these are essentially the same as the Routley star of sets, with the three differences: (1) Having the group $\{0\}$ in place of $\{\}$; this ensures that these maps from groups will always end in groups, not sets. (2) Having $-$ in place of $\neg$; this ensures that it is group operations we are dealing with, not logic operations. (3) Having G and N instead of languages of theories; but the latter stipulates a kind of syntax all the same. Combined, we have an appropriate generalisation.

The aim of the following results is to show that the Routley star on groups is appropriately functorial for groups. We proceed in a somewhat roundabout way.

We recall results about factor algebras. In the rest of the chapter, it is convenient to switch to quotient algbra notation $G \ / \ N$ instead of $G - N$, so as to use the minus for set difference. Given any group G and normal subgroup N, there exists a homomorphism h onto the quotient algebra $h : G \to G \ / \ N$. In particular h takes the elements of its kernel N to 0, and $h(G - N) = G/N - \{0\}$.

This h is exactly what we wish to identify with the Routley star. In short, we show that homomorphisms furnish models for the Routley star. Because of that, too, we can "atomise" the behaviour of the Routley star, so that its behaviour on groups is determined by its behaviour on individual members. We thus proceed by defining (1) the star of an individual element, as its homomorphic image, then (2) extend the star to groups in the natural way, then (3) show that star so defined satisfies the above definition of the Routley star.

Definition 43 $* := h$.

We can immediately say:

$x \in G$ implies $x^* = h(x) \in G \ / \ N$

$x \in G - N$ implies $x^* \in G \ / \ N - \{0\}$

$x \in N$ implies $x^* \in \{0\}$ that is $x^* = 0$.

Proceeding to build up the star of a group by the stars of its elements:

Definition 44 *For any group G, $G^* := \{x^* : x \in G\}$.*

It will be useful to register that changing bound variables $G^* = \{y^* : y \in G\}$. It also follows that: $x^* \in G^*$ iff $x \in G$.

7.4. AN APPLICATION TO HOMOLOGY

We now want to show that with these identifications, the star satisfies the condition for the Routley star above. This proceeds by two lemmas, one showing that $N^* = \{0\}$, and the other showing that this implies that $N^* = \{x : x \in N \ \& \ -x \notin N - \{0\}\}$. The fact that N^* as modelled by the homomorphism really does satisfy the condition for the Routley functor then follows.

Lemma 45 $N^* = \{0\}$.

Proof. In two parts: (a) $0 \in N^*$, and (b) 0 is the only $x \in N^*$. Part (a): From above, $0^* \in N^*$ iff $0 \in N$. Hence $0^* \in N^*$. But $0 = h(0) = 0^*$. So $0 \in N^*$. Part (b). It suffices to prove that if $x \in N^*$ then $x = 0$. If $x \in N^*$, then from $N^* = \{y^* : y \in N\}$ we have that for some y, $x = y^*$ and $y \in N$. That is, $x = h(y)$ and $y \in N$. But from $y \in N$, $y^* = h(y) = 0$. But $x = y^*$, so $x = 0$. ∎

Lemma 46 $N^* = \{0\}$ implies $N^* = \{x : x \in N \ \& \ -x \notin N - \{0\}\}$.

Proof. From $N^* = 0$, we have, for all x, $x \in N^*$ iff $x = 0$. If $x = 0$ then certainly $x \in N \ \& \ -x \notin N - \{0\}$. But conversely, if $x \in N \ \& \ -x \notin N - \{0\}$, then since $x \in N$ iff $-x \in N$, we have $-x \in N \ \& \ -x \notin N - \{0\}$, hence $-x = 0$, hence $x = 0$. That is, $x = 0$ iff $x \in N \ \& \ -x \notin N - \{0\}$. Hence $x \in N^*$ iff $x \in N \ \& \ -x \notin N - \{0\}$. Hence $N^* = \{x : x \in N \ \& \ -x \notin N - \{0\}\}$. ∎

It follows from these two lemmas that the homomorphism of an arbitrary quotient group satisfies the conditions to define the Routley functor on groups. In short, we can further conclude that, under the above homomorphism:

Theorem 47 $G^* = G \ / \ N$.

Proof. Follows from $h(G) = G \ / \ N$. ∎

7.4 An Application to Homology

It is apparent that the above amounts to a redescription of the theory of factor groups in terms of the Routley star. Factor groups are everywhere that an identification is sought. Thus in that sense we have nothing special. But in another sense, it enables the Routley star, as determined by the local homomorphisms, to take on the local character. For example, applied to boundary homomorphisms and their associated homology groups, but not generally elsewhere, we have:

Theorem 48 $*_n *_{n+1} = 0$.

Proof. Follows immediately from the central result of homomogy theory: $\partial_n \partial_{n+1} = 0$. ∎

And also:

Theorem 49 $Ker(\partial_1)^*$ *is the first homology group,* $Ker(\partial_n)^*$ *is the n-th homology group.*

Proof. Starring $Ker(\partial_1)$ is equivalent to forming the factor group under the boundary homomorphism whose kernel is the image of ∂_2, which is how the first homology group is defined, and similarly for higher n. ∎

We also notice that in this construction, the Routley functor commutes with homology operators, forming exact sequences and the like.

*

It is clear that the Routley functor in these different contexts is a single thing: the difference between its effect on groups and its effect on sets of sentences is due to the difference between the contents of the groups and the contents of the sets of sentences. A set of sentences such as $\{p, \neg p, q, r\}$ under star has the p removed and everything else left alone. But groups invariably contain a negative $-x$ for every x, so groups when starred have everything wiped out. This effect can be obtained for sets of sentences when they are deductively closed (theories), depending on the properties of the background logic, for example, whether the logic has the law of Double Negation DN. Neither open set negation nor closed set negation obey full DN, but many logics do, such as the Anderson-Belnap family of relevant logics. If DN holds then the theory which is the deductive closure of $\{p, \neg p, q, r\}$ contains $\neg^n p$ for every $n \geqslant 0$. Starring wipes out all the $\neg^n p$, but leaves q and r alone. In sum, for theories of such logics, starring makes inconsistency into incompleteness, and vice versa (proved in detail in Routley and Routley 1972). In the next part of the book we will make use of such properties of starring, so it is as well to signal ahead of time a shift in background logic to DN, at appropriate points in Part Two. Without the consistent presence of propositions such as $\{q, r\}$, we have a kind of reverse of the above modelling of the behaviour of the Routley functor by means of boundary homomorphisms, namely a kernel-like collapse under the action of the star, to the null linguistic theory.

*

7.4. AN APPLICATION TO HOMOLOGY

This completes our discussion of inconsistent aspects of the homology theory of boundaries. Needless to say, we have only scratched the surface of a large topic. One direction it could be pursued is to extend the functorial action of the Routley functor to various maps of chain complexes, especially exact sequences. Here category theory would be appropriate, in keeping with the origin of that subject in algebraic topology. Another direction is the concept of homotopy. Homotopy theory classifies figures in a more fine-grained way than homology theory - if two maps are homotopic they are homologous, but not in general vice versa - while conversely homology looks at the broader picture as it were. Yet another direction is the subject of cohomology, which deals with a kind of dual or inverse action to homology.

However, we will take a different turn. The example of "impossible pictures" cries out for treatment by the theory of inconsistency. So in the remainder of this book, we will see what sense can be made of inconsistent images, by applying the techniques of the theory of inconsistency to the problem.

Figure 7.2: Pyramid with perspective

Part II

Inconsistent Images

Chapter 8

Impossible Pictures

8.1 Introduction

For the remainder of this book, we will be concerned with making mathematical sense of the so-called "impossible pictures", or more accurately, *inconsistent images*. To forestall confusion, the sense of "impossible" here means real pictures whose content is of logically impossible or contradictory objects. Describing the project like this implies that there is a significant task ahead. It is not enough for an image to *look* impossible or somehow strange, though of course this is an important datum. Impossible images must be *shown* inconsistent, by providing enough logical analysis to display a proposition A and its negation $\neg A$ as part of the content of the image. At the same time, there must be enough structure in the analysis to give a sense of comparability with the successful analysis of figures within geometry. As we will see, these desiderata can be satisfied in at least some cases, but remain problematic for others.

It is worth being clear on what inconsistent images are *not*. They are not images of something which is merely physically impossible. A picture of a man the size of the Empire State Building is a picture of something which is doubtless impossible because of the impossibility of human flesh growing to that size, but it is not a contradictory state of affairs. A picture of Superman flying is not inconsistent; but no-one can fly unaided, we lack the bodily means to do so.

More importantly, inconsistent images are not *ambiguous* images, that is images which are susceptible of more than one interpretation, such as the well-known duck-rabbit, or candlestick-faces. Such images allow switching aspects, but there is no contradiction. I am inclined to treat them not as inconsistent, but as its dual, incomplete, or more strictly, *non-prime*. Non-primeness is a kind of generalisation of incompleteness: a theory is *prime* when for each disjunction it contains, it also contains one or both of the disjuncts, so non-primeness is containing some disjunction while lacking both disjuncts. Similarly, an image is ambiguous when it permits a disjunctive content (duck *or* rabbit), but does not have enough determinate information to force the viewer to settle on one disjunct for its content. That is why we are able to switch readily from one "interpretation" to the other, because the content permits more than one precisification. The content shorn of interpretation is less than either, it is disjunctive while lacking each of the disjuncts. A qualification should be registered: it is possible for an image to be *both* contradictory *and* ambiguous, indeed *ambiguously contradictory*, that is having a disjunctive content where the disjuncts are both inconsistent. We will not take this up here.

For a sense of the difficulty of classifying inconsistent images theoretically, we need to have a grasp of the very large and diverse variety of inconsistent images that have been constructed, particularly in the twentieth century. For a good overview, see the survey by Bruno Ernst (1986). But the story begins well before the twentieth century. We have been able to trace them back to the walls of Pompeii (as Michael Newell pointed out to us). There were also a few examples in medieval alterpieces, where it seems the virgin and child could not be obscured by the columns in the pictures, and so were placed impossibly occluding those columns. The great 18th century artist and draftsman Piranesi included some impossibly-oriented stairways in the dungeons of his *Carceri* series. In the twentieth century, Marcel Duchamp drew an impossible bed, and Rene Magritte contributed a horsewoman impossibly occluding, and occluded by, trees in a forest. But inconsistent images came of age with Oscar Reutersvaard. One day in 1934, bored in his high school Latin class in Stockholm, the 17-year old Oscar doodled as in Figure 8.1

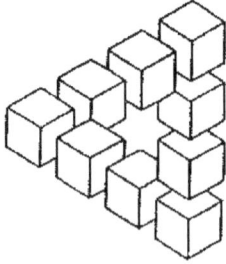

Figure 8.1: Reutersvaard Opus 1

8.2. CONSISTENT MATHEMATICAL APPROACHES

Thereby began a brilliant career in which he drew more than 4,000 images. He was honoured in the 1980s by the Swedish government in a number of stamps featuring his creations.

More than 20 years later, the idea was re-discovered, by the Penroses and M.C.Escher, who at the time appear to have been unaware of Reutersvaard's work. In (1958), L.S.Penrose and his son Roger published a two page paper in the *British Journal of Psychology*, one page of which contained drawings and a photograph. At about the same time, Escher, who was communicating with the Penroses, contributed the masterpieces *Belvedere, Waterfall,* and *Ascending and Descending*. It is worth noting that Escher seems not to have done more than these three that are genuinely inconsistent (as opposed to merely physically impossible, or in other ways strange). In the later part of the twentieth century, under these stimuli, the project of inconsistent images really got going strongly, and names such as Bruno Ernst, Jos de May, Sandro del Prete and Al Seckel have all been prominent, among many others.

8.2 Consistent Mathematical Approaches

Now we briefly describe three "consistentist" approaches to giving a mathematical description of at least some inconsistent images. We will see that they share a common limitation.

First, we have Thaddeus Cowan (1974). Cowan focussed on four-sided figures with a hole in the middle, though the analysis applies to any figure with three or more straight sides around the hole. Examples are in Figure 8.2.

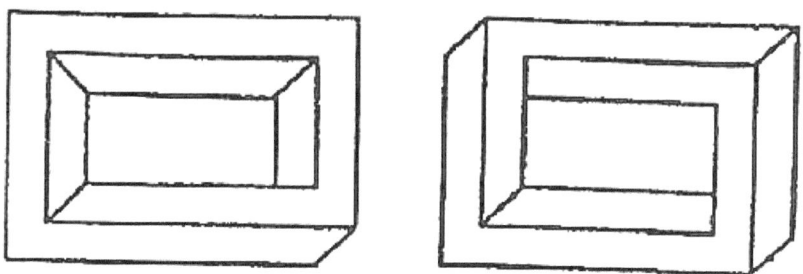

Figure 8.2: Inconsistent four-sided figures

Cowan was able to classify 2-dimensional corner elements in terms of the 3-dimensional elements they represent, using the concepts front, back, inside and outside. This results in four kinds of corners (Figure

8.3). With four corners, there are 256 different figures. Applying the theory of the *braid group*, Cowan obtained necessary and sufficient conditions for four corners to assemble into a possible object. Essentially, a picture is impossible if it corresponds to a non-trivial braid around the hole.

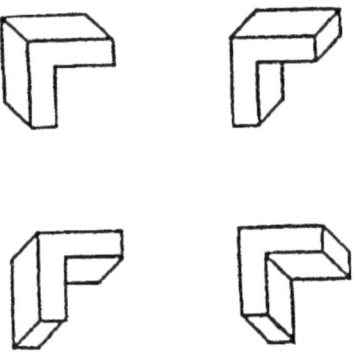

Figure 8.3: Corner elements

This is a very excellent and useful theory. However, it leaves something out. In effect, it is telling us that if the conditions fail, then this is not a picture of an impossible object. Now that is a perfectly consistent state of affairs. It fails to capture our sense of what it is a picture *of,* its *content*, namely of an object with impossible properties, which render it unable to exist.

Then we have a contribution by George Francis (1987). Francis asked the sensible question: is there a consistent 3-D space in which a figure like the Reutersvaard-Penrose triangle can live? This immediately gives a positive heuristic to the problem. Francis pointed out that the triangle could exist in a certain 3-D non-Euclidean space, specifically $R^2 \times S^1$. It is difficult to represent this on the page, but the case $R^1 \times S^1$ as embedded in R^3 is familiar to us as the cylinder (Figure 8.4).

The triangle, and for that matter the four-sided figures above, are thus "wound around" the rolled-up S^1 dimension, and so are able to join back on themselves, as they appear to. Francis is clearly right here. But there is another way to view his point. Why does the triangle appear to be impossible, if it is really possible? Clearly because we are hard-wired to think of the space we live in as R^3, not $R^2 \times S^1$. Our sense of impossibility is produced by a clash between what presents itself, and our expectations about the space nearby.

Well and good, but we can go further, because this gives us a handle

8.2. CONSISTENT MATHEMATICAL APPROACHES 73

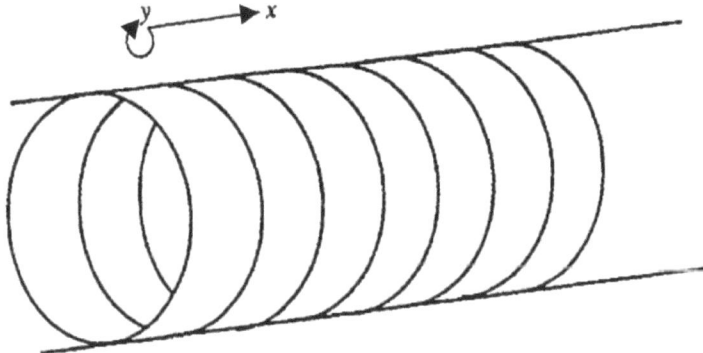

Figure 8.4: $R^1 \times S^1$

on the problem of the content of the experience of impossibility. The content is an amalgam of the projection the figure presents itself to have, and the incompatible expectations about the nearby space. *How is it that the triangle seems*? It seems to be something joined-up with a hole in the middle, something topologically similar to a torus, in short. This content is what leads us to say "I see it but it is impossible". The mind is thus projecting something onto its perceptions, the expectation that space is R^3. No consistent object behaves like this, but an inconsistent object can, *and that requires an inconsistent theory to describe it*.

We will look more closely at this in the later chapter on the triangle. But we should turn now to Roger Penrose. In (1991), he made a further significant advance when he described the situation using cohomology theory. The idea was broadly similar, namely that a picture of an apparently impossible object can indeed represent a consistent object, but one with different connectedness. For example, one can imagine that that the triangle is a picture of 3 disassembled parts lying at different distances from the viewer, but lined up so that they look joined up. That is, they are lined up so that they project down onto the 2-D image. Now if the image is consistent, then it is possible to re-assemble the pieces into an object with a central hole, whereas if the image is inconsistent then this re-assembly is not possible. Penrose shows how to describe this using cohomology groups.

This is clearly a useful advance. But again one feels that something has been left out. The mind doesn't think "here are 3 disassembled parts", its content is of something assembled. Nor does it simply "fail to dis-identify" separate points which are in a straight line with the eye. It is of course possible to imagine the content as a dis-assembled object, but there remains the strong percept of an assembled, but impossible object. It has a *positive* content. The mind completes the picture, it seems to have a default mode where it actively identifies points that it cannot distinguish. But to do that, the mind's contents must be represented by an inconsistent theory.

8.3 Conclusion

It is worth comparison with projective geometry. Perspective presents us with an obviously consistent perception: the railway lines *look* as if they meet. This is an objective phenomenon; if you take a photograph the lines will meet at a point on its surface. Perspective does not *look* paradoxical. What we know, is that the lines do not meet in 3-D; but this is cognition at a higher level. It is not part of the content of the percept, it is not "projected onto the percept" as it were. In this respect the content differs from that of the inconsistent images, whose look is paradoxical. In each case, the content is a collection of propositions which form a mathematical theory. Why we are interested in such theories, however, is because of the presence of the human perceptual apparatus. Thus, the present account is a thoroughly *cognitive* justification of the application of inconsistent theories. There is no suggestion that there really exists an inconsistent object. Even so, we have an entirely appropriate use of paraconsistent methods, which falls under the *epistemological* justification of paraconsistency (see *e.g. Inconsistent Mathematics* 1995 Chap 1).

Faced with the rich variety of inconsistent images that have been drawn, how are we to bring enough order into them even to begin mathematical classification? For many years, I held the conjecture that *all inconsistent images are occlusion paradoxes*. An occlusion paradox is defined as an inconsistent image which can be rendered consistent by the reversal of one or more occlusions. Now *some* inconsistent images are certainly occlusion paradoxes, for example the inconsistent Necker cubes below, or the Reutersvaard-Penrose triangle below, or multiply-sided variants thereof, such as Cowan's. But the conjecture does not seem to fit at least two broad classes of images. These are instanced by the two images on the bottom row in Figure 8.5: the Ernst stairs, and the Schuster fork.

In the absence of anything better therefore, I propose a fourfold classification as in Figure 8.5: the cube, the triangle, the stairs and the fork.

In the succeeding chapters, we will deal with these in order. Perhaps

8.3. CONCLUSION

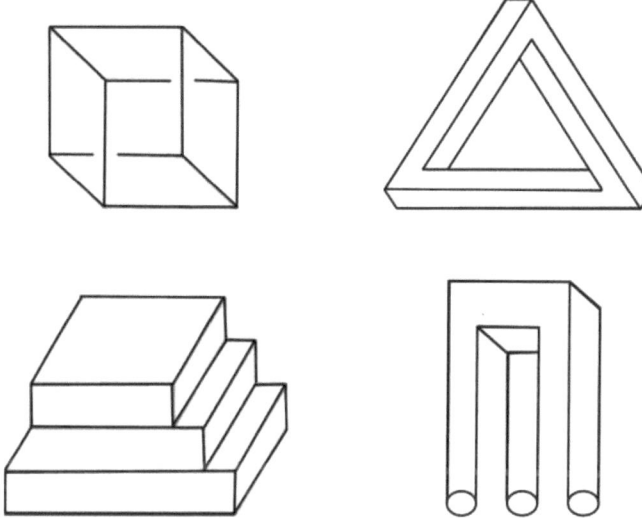

Figure 8.5: Four forms

the classification falls short of what contradictory images there can be. If we can determine this, then we will have learned something important about the kind of inconsistencies there are. We will have learned more about what structures the inconsistent contains.

Figure 8.6: Wrong right angles

Chapter 9

Logical Analysis of Necker Cubes

9.1 Introduction

In this chapter, we begin an analysis of Necker cubes, which will occupy us for several chapters. The cubes will be inconsistent, incomplete, both or neither. The present chapter concerns itself with an analysis at the level of logical theories. This is essential if it is to be shown that the content of these images represents a genuine inconsistency, an A and a $\neg A$. The following chapters will apply linear algebra to the problem.

We begin by noticing an impossible image, which I will call "Escher's cube" (Figure 9.1).

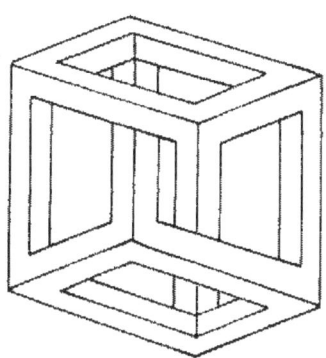

Figure 9.1: Escher's cube

Readers familiar with the works of M.C.Escher will recall that this appears in the hands of the little man in the corner of Escher's *Belvedere*. Bruno Ernst (1986, 74) credits Escher as the first to draw this particular kind of image, in 1957-8. Since Ernst was well aware of the legitimacy of Oscar Reutersvaard's claims to primacy with so many inconsistent images, this is a good reason to credit Escher here. Readers will also notice a similarity between Escher's cube and the image called "Cochran's cube", or "the crazy crate" (Figure 9.2). But they are not the same; the wire-framed version of Escher's cube is inconsistent (Figure 9.6), whereas the inconsistency in Cochran's crate lies elsewhere, as its wire frame is consistent.

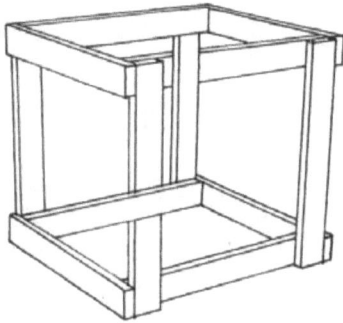

Figure 9.2: Cochran's cube

Now Escher's cube obviously invites comparison with the usual Necker "cube" (Figure 9.3). This is placed initially in scare quotes to emphasise a point which will become apparent as we proceed, namely that it is more accurately called a Necker cube *diagram*.

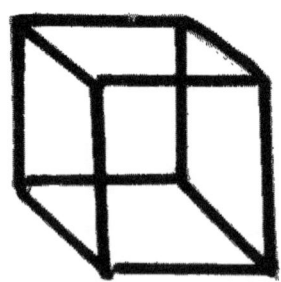

Figure 9.3: The usual Necker cube

9.1. INTRODUCTION

The usual story of the Necker cube is that it is an *ambiguous* image, in the sense of ambiguity described in the previous chapter. The viewer's *gestalt* switches between two aspects, one with one face in front, then the other. This story will be altered and extended as we proceed. The two ways to "disambiguate" the Necker can be displayed by introducing occlusions at the two internal crossings, in the manner of the convention governing knot diagrams in knot theory (Figure 9.4).

 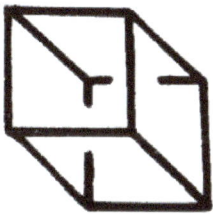

Figure 9.4: Two disambiguated Neckers with occlusions

This can be made even more apparent by bringing in colours as well: red lines for one face, blue lines for its opposite face, and green for connecting lines (Figure 9.5).

Figure 9.5: Disambiguated Neckers with occlusions and colour

But now notice something important: the device of occlusion (with the help of colour) provides for two impossible cubes, where one of the two crossings favours the red, while the other crossing favours the blue. Schematically drawn, these are Escher's cube and its mirror image (Figure 9.6).

This chapter and the next two aim to explain the experience of Escher's cube and its cognates using a variety of tools, by defining inconsistent theories which describe the content of the experience. In the chapter following these, we will see how to extend the description to chains of Neckers displaying higher-order inconsistency.

CHAPTER 9. LOGICAL ANALYSIS OF NECKER CUBES

Figure 9.6: Impossible Neckers: the Escher cubes

9.2 Level 1 Analysis: Two Dimensions

We begin with the characteristics of the image, namely the 2-D collection of lines and points at which they cross. This amounts to all you can see without reading a third dimension into your perceptual construct. There are 8 points and 12 lines. In saying that there are 12 lines, I am including as a single line of the same colour those lines that are broken in the diagrams with crossings. This way of talking makes for some simplicity when we move up a dimension to 3-D.

We proceed to present a set of axiomatic theories. The background logic is left underdetermined, but is thought of as paraconsistent and paracomplete, such as a suitable 4-valued logic which sustains non-trivial inconsistent or incomplete theories. A key requirement is that it contains the law of Double Negation. Perhaps the simplest suitable background logic would be Anderson-Belnap first degree entailment FDE (see for example Routley and Routley (1972), or Priest (2001)). FDE is the constant first degree part of many natural paraconsistent logics, and satisfies the conditions for the Routley functor to be used. As will become clear, the interesting facts about these Necker cubes and their classifications are all at the first degree level: atomic relations compounded with (&, ∨, ¬), and entailments → between them. Quantifiers (∃, ∀) can be eliminated in favour of disjunctions and conjunctions because these geometrical objects contain only a finite number of elements (vertices, edges, faces, crossings). Higher-degree nestings of entailments do not seem to reflect themselves in the images, hence logicians' disputes over the properties of nested entailments are not relevant here. This is in line with the observation made in an earlier chapter, that inconsistent mathematics is invariant over a large class of background logics.

On the other hand, it is worth noting that not all paraconsistent techniques are equally suitable. For one thing, non-adjunctive techniques do not seem to be applicable here, though they may have application for

9.2. LEVEL 1 ANALYSIS: TWO DIMENSIONS

other kinds of inconsistent images, for example the fork. For another thing, FDE negation is none of the topologically-based negations of the first part of this book. Its distinguishing feature, as the Routleys showed (1972), is that the star of an inconsistent theory is an incomplete theory, and vice versa, and the star of a consistent theory is a complete theory and vice versa, so that a consistent complete theory is a fixed point under the action of star. This makes for a different sort of duality, which is particularly suitable in this chapter and the next three.

We focus on the two crossings of the cube as the most important part of the diagram, since it is by changing what is happening there that enables a switch from inconsistent to consistent images and back again. So fix and describe the whole diagram but for the two crossings. Call the crossings $C1$ and $C2$. At each crossing can be placed a red point RP, a blue point BP, both or neither. This can be likened to dropping paint of the appropriate colour at the crossings. The basic relation is **x@C**, where **x** is a coloured point and **C** a crossing. Thus for crossing $C1$, there are 4 possible theories:

Ax 1: $RP@C1$ (but lacking $BP@C1$)

Ax.2: $BP@C1$ (but lacking $RP@C1$)

Ax.3: Both $RP@C1$ and $BP@C1$

Ax.4: Neither $RP@C1$ nor $BP@C1$

A point to note for the logician is that these are not quite "axioms" in the traditional sense, since they involve exclusions as well. But this is no real problem, one can either employ an exclusion operator in the metalanguage, or simply take Ax1-4 not as axioms, but as more general stipulations of the theory of interest.

The four axioms also describe crossing $C2$ as well. Hence there are 16 possible Necker diagrams which can be built out of these elements (Figure 9.7).

It might be thought to be problematic that there can be both red and blue at a crossing at once. After all, this might be thought to violate the *natural* principle, a favourite with philosophers, that a thing cannot be both red and blue all over at the same time. However, at the present level of analysis, there is nothing obviously inconsistent about placing both red paint and blue paint at a crossing at the same time. Various things might happen, so no assumption should be made as to extra-logical principles.

Having said that, I wish to signal endorsement of this natural principle. This can be incorporated as an *axiom of local consistency* of the @ relation:

CHAPTER 9. LOGICAL ANALYSIS OF NECKER CUBES

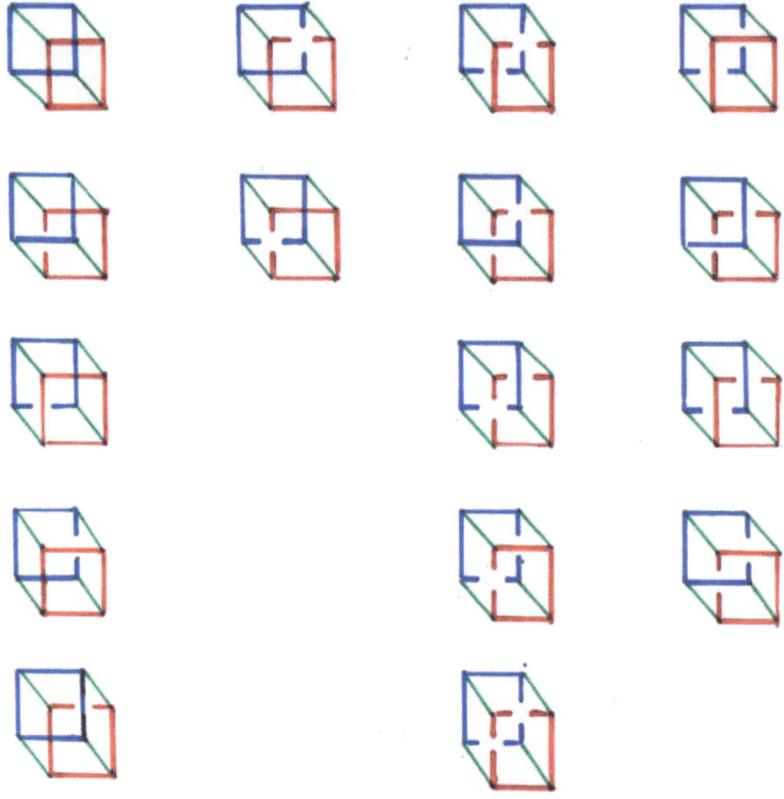

Figure 9.7: Sixteen Necker cubes

Ax. LoCon@:

$$(C)(RP@C \rightarrow \neg BP@C)$$

This axiom is routinely satisfied by axioms Ax.1,2,4. It is evident that in the presence of Ax.3, an inconsistent theory is generated. These facts can be expressed with some definitions:

Definition 50 *(Local consistency and completeness with respect to @). A theory Th is locally consistent at C w.r.t @, iff for each colour x, not both $x@C$ and $\neg x@C$ are in Th. Th is locally consistent w.r.t. @ iff for all C, it is locally consistent at C w.r.t @. A theory is locally complete at C w.r.t. @, iff for each colour x, either $x@C$ or $\neg x@C$ are in Th. Th is locally complete w.r.t.@ iff for all C, it is locally complete at C w.r.t. @.*

9.2. LEVEL 1 ANALYSIS: TWO DIMENSIONS

We can drop the "@" when it is obvious, as here, which relation is being discussed. We can now say that Ax.1&2 correspond to locally consistent and complete theories. Ax.3 corresponds to a locally inconsistent theory and Ax.4 to a locally incomplete theory.

Beginning with the two examples of the traditional Necker cube and Escher's cube, we have now seen that our analysis expands the field of possibilities for Necker cubes to 16. We can think of the diagrams as "representations" of locally inconsistent and incomplete theories, an extension of the convention from knot theory that a broken line represents a locally consistent and complete situation. In terms of the above diagram (Figure 9.7) of the 16 Neckers, it can be seen that:

(1) First column. These theories are locally inconsistent and locally complete. The top diagram in this column (the traditional Necker) is trivial, in that all atomic sentences hold, that is Ax.3 holds for both crossings. All the others in the first column have one inconsistent and one consistent and complete crossing, that is Ax.3 holds at one crossing, and either Ax.1 or Ax.2 holds at the other.

(2) Second column. These are locally inconsistent and locally incomplete. That is, Ax.3 holds at one crossing, and Ax.4 holds at the other.

(3) Third column. These are locally consistent and locally incomplete. That is, Ax.4 holds for one or both crossings, and Ax.1 or Ax.2 holds where Ax.4 doesn't. The bottom diagram in this column has Ax.4 holding at both crossings, that is the theory of this diagram is locally incomplete at both crossings. All the others in the third column are consistent and incomplete at one crossing, and consistent and complete at the other.

(4) Fourth column. These are all locally consistent and locally complete. The top two are the classical disambiguations of the traditional Necker, where Ax.1 holds at both crossings, or Ax.2 does. The bottom two are the inconsistent Escher cubes: Ax.1 holds at one crossing, and Ax.2 holds at the other.

The evident duality between Ax.3 and Ax.4 can be given an explanation in terms of the Routley functor, as we see in the next section. This is yet another testimony to the theoretical usefulness of that notion.

We conclude this section by noting that the inconsistency of the two Escher cubes has **not** yet been given an explanation in terms of an inconsistent theory, though there is in place an appropriate theoretical framework for doing so. So far, the Escher cubes are simply locally consistent and complete. But only if an inconsistent theory is forthcoming, can it be said that the sense that we have here of the Escher cubes, that they are somehow contradictory, has been justified. The following sections are devoted to this task.

9.3 Level 2 Analysis: Three Dimensions

We move to the question of what 3-D objects these 2-D diagrams represent. We are mentally constructing a 3-D object out of 2-D data, in the usual fashion of depth perception (one eye). To mark this change of viewpoint, we introduce the common terminology of vertices, edges and faces. Thus, all 8 points on the diagrams become 8 vertices, all 12 lines become 12 edges. For the present secion, we hold off on introducing faces, which come in at the next level.

Crossings are seen as a 2-D artifact of viewing a 3-D object from a particular point of view, that is as devices for registering which of two points in 3-D space that project to the same point in 2-D, is closer, and which is further away. Again, this is in line with the corresponding treatment of 2-D knot diagrams as projections of knots in 3-space. However, it is important to see the present project as building up a 3-D object from 2-D data. So we want to preserve the viewing aspect of the picture. Thus the main new relation we want, is that of *occlusion*, which will be written **E1 occ E2**. This is a relation between edges $E1$ and $E2$. In these theories, there are just 4 edges involved in an occlusion, two red and two blue. Hence, where RE is a red edge and BE is a blue edge, then occlusion is defined by:

Definition 51 *(Occlusion)*

$RE\ occ\ BE := (\exists C)(RE\ and\ BE\ cross\ at\ C\ and\ RP@C)$

$BE\ occ\ RE := (\exists C)(RE\ and\ BE\ cross\ at\ C\ and\ BP@C)$.

Thus, while the occlusion relation holds between edges in 3-D, it derives from data in the 2-D diagrams we began with.

It is clear that where Ax.1 or Ax.2 hold for each crossing, this ensures that, for each crossing, just one of the above two conditions for occlusion obtain, again as in knot theory, since we do not have both $RP@C$ and $BP@C$ for either C. When Ax.3 holds at a crossing, both $RE\ occ\ BE$ and $BE\ occ\ RE$ hold there. When Ax.4 holds at a crossing, then neither edge occludes the other there. These can then be read back into the diagram.

9.3. LEVEL 2 ANALYSIS: THREE DIMENSIONS

It then becomes of interest to consider a local consistency axiom for **occ**. This says that if one edge occludes the other then the latter edge does not occlude the former, that is the asymmetry of **occ**. In knot theory this is so standard that it is rarely if ever mentioned.

Ax.LoConOcc:

$$(x, y)(x \; occ \; y \to \neg y \; occ \; x)$$

This axiom expresses the consistent state of affairs when one edge occludes the other at their crossing. Again, it is to be endorsed and incorporated here. For diagrams where there is just one colour at the crossing point between the two edges, LoConOcc is satisfied, as it is if Ax.4 holds. When Ax.3 holds, we have a theory that is locally inconsistent w.r.t. the **occ** relation. That is, LoConOcc is satisfied if we treat a crossing where both colours are present as inconsistent, since the presence of both $RE1 \; occ \; BE2$ and $BE2 \; occ \; RE1$ ensures further that $\neg(RE1 \; occ \; BE2)$ and $\neg(BE2 \; occ \; RE1)$.

Summing up:

Theorem 52 *If LoConOcc is added to Ax.1-4, there are the following consequences:*

For Ax.1&2, the theory of occlusion is both consistent and complete
For Ax.3, the theory of occlusion is locally inconsistent and complete
For Ax.4, the theory of occlusion is locally consistent and incomplete.

The relations between these four axioms can be described using the Routley functor. If we write (Ax.1) for the zero degree (@, occ) theory generated from Ax.1, LoCon@ and LoConOcc, we have:

Theorem 53
$(Ax.1)^* = (Ax.1)$
$(Ax.2)^* = (Ax.2)$
$(Ax.3)^* = (Ax.4)$
$(Ax.4)^* = (Ax.3)$.

Proof. (Ax.1) and (Ax.2) are consistent and complete, so the star leaves them untouched. If Ax.3 holds at C, then we have seen that all of $RP@C, BP@C, RE1 \; occ \; BE2$, and $BE2 \; occ \; RE1$ and their denials hold. Applying the star removes contradictory pairs, and that is the condition for (Ax.4). Similarly for Ax.4, since the effect of the star on missing pairs $(A, \neg A)$ is to replace them. ∎

In terms of our diagram of the 16 Neckers, the members of the 1st column are *-duals of the 3rd column, the members of the 2nd column are *-duals of each other, and the members of the 4th column are self-dual.

It is the 4th column of 4 members that we are interested in, since it contains the two Escher cubes. So far all we can say is that they are locally consistent and complete, yet they are somehow inconsistent. How can this be? The answer lies in inconsistency in a more global sense, as we see in the next section.

9.4 Level 3: Faces

We add faces to the theory. A face (for the cube) is a figure having 4 edges intersecting at the corners in 4 vertices. The usual consistent theory is that cubes have 6 faces. Continuing with our colour scheme for the edges, there is one face with all red edges, one with all blue edges, and four faces each with two green edges. As we have drawn them, there are no occlusions invloving green edges, so we focus on just the red and blue faces, RF and BF.

We define the relation **in front of** for faces. This relation obtains only between the red and the blue faces. Taking a lead from the consistent case, we define:

Definition 54 *(in front of)*
$F1$ *in front of* $F2 :=$ *some edge of* $F1$ *occludes some edge of* $F2$.

In the case of the two "normal" Neckers (top two, 4th column), we have that RF is in front of BF, or vice versa, but not both. Now it might be thought that we should have required that *all* edges of $F1$ occlude all edges of $F2$. However, this is unnecessary in the normal cases. Independently, if one tries out the cognitive experiment of covering one of the crossings and asking which colour occludes which at that crossing, the natural answer to give is that it is the same as at the visible crossing. (I owe this observation to Lloyd Humberstone), Thus, the definition is built into our expectations.

Now it is appropriate to consider an *axiom of global consistency* for the relation **in front of**.

(Ax.GlobCon)
$$(x,y)(\ x \text{ in front of } y \rightarrow \neg y \text{ in front of } x)$$

This asymmetry of **in front of** guarantees, for the two Escher cubes, that we have: RF *in front of* BF and BF *in front of* RF and their denials.

Introducing a definition:

9.4. LEVEL 3: FACES

Definition 55 *(Global Inconsistency)*

*The theory of a Necker is globally inconsistent iff it is inconsistent with respect to the relation **in front of**, otherwise globally consistent.*

It follows that the two Escher cubes are locally consistent and complete, but globally inconsistent. I submit that this is independently reasonable as an expression of our intuition that the Escher cubes are impossible. From the classical perspective they are "defective of face making". But from the point of view of inconsistent mathematics they have inconsistent faces, that is faces with inconsistent properties, because faces end up being inconsistently in front of one another.

*

This chapter has been concerned to analyse the Escher cubes in terms of logical theories. Cochran's cube has not been treated as it requires a different story. But there is more to be said about the structures above in this chapter. It turns out that linear algebra is useful in describing them. In particular, linear algebra permits generalisation to complexes of cubes. So we turn to that.

Figure 9.8: Observatory

Chapter 10

Linear Algebra and Necker Cubes

(co-authored with Steve Leishman)

10.1 Introduction

In the previous chapter we began an analysis of the two Escher cubes, We found a class of Necker cube variants, sixteen in number. Some of these are locally inconsistent, some locally incomplete, some both and some neither. We were able to define a sense of global inconsistency that distinguishes the Escher cubes from their variants. In this chapter, we begin an account of these structures using linear algebra. In two following chapters this is extended to the Routley functor, then to chains of Neckers displaying a certain higher-order inconsistency.

10.2 Primary and Secondary Matrices

In the previous chapter, faces and edges were coloured red (R) and blue (B) respectively (with connecting edges coloured green). Each Necker diagram has two places where red crosses over (occludes) blue or vice versa. These are termed crossings, $C1$ (left hand crossing) and $C2$ (right hand crossing). The Neckers differ from one another in just two places, hence we ignore all other data on the images as constant and therefore irrelevant to their differences. The presence or absence of R and B at the crossings $C1$ and $C1$ completely distinguishes all sixteen. Thus for each cube there is a pair of simultaneous linear equations:

$$a.R + b.B = C1$$

$$c.R + d.B = C2$$

where each of a, b, c, d are in $\{0,1\}$. For example, "$1.R + 0.B = C1$" can be read "1 of red plus 0 of blue make up crossing $C1$". In turn, the simultaneous linear equations can be written in the usual fashion as a matrix multiplication over the field Z_2, that is Z mod 2:

$$\begin{bmatrix} a & b \\ c & d \end{bmatrix} \times \begin{bmatrix} R \\ B \end{bmatrix} = \begin{bmatrix} C1 \\ C2 \end{bmatrix}$$

Here $\times$ stands for matrix multiplication. We call this the **primary equation** of the cube, and the left hand 2×2 matrix the **primary matrix**. If M is a primary matrix, we also denote it as M_p. Given a matrix designated as a primary matrix, the Necker figure can obviously be recovered.

In the last chapter, the two Escher cubes are described as *locally consistent* but *globally inconsistent*, as exemplified by a *failure of (consistent) face building*. The status of faces in these figures can also be represented by linear equations over Z_2. We write:

$$a'.C1 + b'.C2 = RF$$

$$c'.C1 + d'.C2 = BF$$

where as before the (primed) coefficients are from $\{0,1\}$. For example, "$1.C1 + 0.C2 = RF$" can be read as "The red face is made up by having red at $C1$ and not having red at $C2$". As before, these equations can be written in matrix form:

$$\begin{bmatrix} a' & b' \\ c' & d' \end{bmatrix} \times \begin{bmatrix} C1 \\ C2 \end{bmatrix} = \begin{bmatrix} RF \\ BF \end{bmatrix}$$

We call this the **secondary equation** of the corresponding Necker, and the left hand 2×2 matrix its **secondary matrix**, denoted by M_s.

In linear algebra, the **transpose** of a matrix M, written M^T, is the matrix made up by exchanging rows for columns and vice versa. That is, if a_{ij} is the element in M in the i-th row and j-th column, and a'_{ij} is the element of M^T in the i-th row and j-th column, then $a'_{ij} = a_{ji}$. M^T is the reflection of M across the main diagonal. Evidently, the operation T is an involution, that is $M^{TT} = M$.

Theorem 56 *Let N be any Necker cube with primary matrix M_p and secondary matrix M_s Then $M_s = M_p^T$.*

10.2. PRIMARY AND SECONDARY MATRICES

Proof. From the simultaneous equations, we can deduce that the red face is made up of contributions of $a.R$ at $C1$ and $c.R$ at $C2$. Thus, $a.C1 + c.C2 = RF$. The other case is similar, giving $b.C1 + d.C2 = BF$. Assembling the two equations into the secondary matrix gives the result. (Alternatively, one may prove this result by enumeration of cases and inspection thereof.) ∎

It is recalled from the previous chapter that a Necker cube is classified as **locally inconsistent** if it has both colours, R and B, at one or both crossings. It is apparent that this corresponds to having all 1s in either row of its primary matrix. That is, a Necker is locally consistent just in case no row of M_p is all 1s. The special case where both rows are all 1s is the usual Necker cube familiar from psychology texts. Corresponding to its theory being the trivial theory, we call its matrix the **trivial** matrix. The locally inconsistent Neckers are the left hand column of the diagram; the trivial Necker being the topmost.

A Necker was further classified as (locally) **incomplete** if one or both crossings have no colours present. It is apparent that this corresponds to having all 0s in either or both rows of M_p. The case where both rows are all 0s corresponds to the theory having no (non-logical) statements holding. The incomplete Neckers are the two middle columns of the diagram. Thus, a Necker is complete if no row of M_p is all 0s.

Of the four locally consistent and complete Neckers (the right hand column), the bottom two, the Escher cubes, are inconsistent in a further global sense. As noted before, this was explained as a failure of consistent face building. A consistent face is built just in case its colour is present at both crossings, or absent at both crossings, corresponding to being unambiguously the front face, or unambiguously the back face. It is apparent, therefore, that a face is built just in case its corresponding row in the secondary matrix is all 1s, or all 0s. So, for these four locally consistent and complete Neckers, to be globally consistent is to have both faces consistently; that is no row on M_s is all 1s and no row is all 0s. Since $M_s = M_p^T$, it follows that no column on M_p is all 1s and no column is all 0s. These are the top two Neckers of the fourth column.

On the other hand, since the two Escher cubes also have no row all 1s and no row all 0s, we can isolate them as having respectively one of the two primary matrices:

$$\begin{bmatrix} 1 & 0 \\ 0 & 1 \end{bmatrix} \text{ and } \begin{bmatrix} 0 & 1 \\ 1 & 0 \end{bmatrix}$$

The first of these is the **diagonal** matrix whose leading diagonal is all 1s and there are just 0s elsewhere, and we name it **Id** here. The second is the **antidiagonal** matrix whose antidiagonal is all 1s and there are just

0s elsewhere, and we name it **AntiId** here. We observe, in passing, that in each case, $M = M^T$, and that, in general, $M_p = M_s$ iff $M = M^T$.

We utilise this information in what follows.

10.3 Determinants

We proceed to several "tests" for inconsistency and incompleteness. We are expecially seeking matrix-algebraic conditions which group together the two kinds of inconsistency as being similar. The first obvious test to apply is the nature of the determinant of the matrix. Unfortunately, as we see, while this yields some information, it is a blunt instrument.

The **determinant** of the primary matrix given at the beginning of the last section is the sum $(ad - bc)$, that is the product of main diagonal less the product of the antidiagonal. The determinant of a matrix M is written $Det(M)$. (In $n \times n$ matrices the definition of the determinant is more complicated. The reader also is reminded that the arithmetic being used is that of Z_2)

Theorem 57 *If $Det(M_p) \neq 0$, then M_p is inconsistent.*

Proof. If $ad - bc \neq 0$, then we have two cases. Case (a): $a = d = 1$ and one or both of $b, c = 0$. If one of $b, c = 1$ then we have three 1s in the matrix, so that one row is all 1s, so that the matrix is inconsistent. If both $b, c = 0$, we have the diagonal matrix which was identified above as one of the two Escher cubes. Case (b): $b = c = 1$ and one or both of $a, d = 0$. If one of $a, d = 1$ then we have three 1s in the matrix which is inconsistent. If both $a, d = 0$, we have the antidiagonal matrix AntiId which was identified above as the other Escher cube. ∎

Comment. We note that the converse fails in three cases: the trivial matrix and the two inconsistent and incomplete matrices (a row of 1s and a row of 0s in M_p). Thus it would seem that the determinant is not such a useful instrument in identifying consistency and inconsistency.

10.4 Nullity

A more precise test is given by the concept of nullity. In linear algebra, we have the notion of the null space. The **null space** of an $n \times n$ matrix M is the space of (consistent) solutions of the matrix equation $Mx = \mathbf{0}$, where the unknown x is an n-ary column vector of variables, and $\mathbf{0}$ is the n–ary column vector which is all zeros. The **nullity** of the matrix

10.5. THE STRUCTURE OF THE NULL SPACE

is the dimension (number of independent solutions) of the null space. Note that the matrix equation $Mx = \mathbf{0}$ always has at least one solution, namely all the variables equal to zero. If this is the only solution, it is termed the zero-dimensional solution.

Theorem 58 *Let M_p be non-trivial. Then M_p is complete iff the nullity of M_p is zero.*

Proof. Left to Right: If M_p is complete then no row is all 0s. If M_p is also non-trivial, then not all entries are 1s. So the number of 1s is 2 or 3. If any row is all 1s, then only the solution $(R, B) = (0, 0)$ is possible. If no row is all 1s then M_p is either the diagonal matrix or the antidiagonal matrix. In each case the nullity is zero since the occurrences of 1 force that variable to be zero. Right to Left: If M_p is incomplete then one of both rows are all 0s. Any row of all 0s does not disturb the set of solutions, since everything is a solution to it. Thus the space of solutions is that of the other row. But this must have at least one other solution: if the row is both 1s then $[1, 1]$ is a solution, if the row has one 1 then a zero in that place and anything in the other place suffices for a solution, and if the row is both zeros then anything is a solution. ∎

Comment. From the theorem, a non-zero nullity indicates that there are other ways to satisfy the matrix equation. The nullity can thus be thought of as a measure of the collection of ways in which R and B come together to make an assignment to the crossings $C1$ and $C2$. A non-zero nullity indicates that there are more ways of doing this than simply having no colours anywhere. This in turn is a mark of incompleteness: an incomplete crossing can be completed in more than one way, which indicates more than one solution for $[R, B]$.

However, note too the exception, when M_p is trivial. This is complete and inconsistent. However, $[R, B] = [1, 1]$ is also a solution of the nullity equation, since in Z_2, $1 + 1 = 0$. So the nullity of M_p is non-zero. However, its converse (incompleteness implies non-zero nullity) holds irrespective of triviality.

10.5 The Structure of the Null Space

The above suggestive result invites a related test for inconsistency and incompleteness, namely to look at the structure of the null space of the primary and secondary matrices. In an improvement over the previous sections, the conditions obtained are general.

Theorem 59 *A Necker is inconsistent iff the null space of its primary matrix satisfies R=B.*

Proof. Left to Right: If a Necker is inconsistent then we have either that at least one row is $[1,1]$, or is the diagonal or the antidiagonal. If any row is $[1,1]$ then the only way this can compute to zero is if $R = B = 0$ or $R = B = 1$. Either way, $R = B$. For the diagonal matrix we have one row $1.R + 0.B = 0$, which implies $R = 0$. The other row is $0.R + 1.B = 0$, which implies $B = 0$. Hence $R = B$. The calculation for the antidiagonal matrix is the same. Right to Left: Any zero entry in the primary matrix, and thus the row $[0,0]$, does not disturb the set of solutions The row $[0,1]$ implies that $B = 0$ and R is any, so that $R = 1$ is possible, so that $R \neq B$ is possible. The calculation for the row $[1,0]$ is the same. Hence matrices with three zeros do not satisfy $R = B$. Matrices with three or four ones are all inconsistent. This leaves six. Two of these, the diagonal and the antidiagonal, are globally inconsistent. Two more, having a row of 1s and a row of 0s, are locally inconsistent. The final two, where the two rows are identical, force one colour to be 0 while the other colour can be any, so $R \neq B$ is possible ∎

Comment. The test asks which combinations of R and B make for each crossing to be zero. The inconsistent Neckers are just those for which the primary equation requires $R = B = 0$ or $R = B = 1$. This is perhaps not surprising in that $R = B = 0$ is always a solution, and otherwise only $1.1 + 1.1 = 0$ gives an inconsistent row.

Theorem 60 *A Necker is complete iff the null space of its secondary matrix satisfies $C1 = C2$.*

Proof. Left to Right: A Necker is complete iff no row of its primary matrix is all zeros. Hence iff no column of its secondary matrix is all zeros. Such matrices have at least two ones in them. Matrices with exactly two ones and no column all zeros are (a) the diagonal, (b) the antidiagonal, or (c) one row all ones and the other all zeros. Case (a) gives $1.C1 + 0.C2 = 0$ which forces $C1 = 0$; which combined with $0.C1 + 1.C2 = 0$ also forces $C2 = 0$; whence $C1 = C2$. Case (b) is the same. Case (c): the row of all ones forces $C1 = C2 = 1$ or 0. Otherwise, matrices with at least three ones in them force $C1 = C2$, because as before the row $1.C1 + 1.C2 = 0$ requires $C1 = C2 = 0$ or 1. Right to Left: Conversely, if a Necker is incomplete, then there is some column of the secondary matrix which is all zeros. Without loss of generality, we may assume it is the left column. Columns of the secondary matrix correspond to crossings. So the value of that crossing can be either 0 or 1 and the secondary equation is still satisfied. If the other column is also all zeros then the matrix is all zeros

10.6. THE UNIT EQUATION

and all solutions are possible so that $C1 \neq C2$ is possible. If the other column has at least one 1 in it, then that column is forced to be 0, since $0.C1 + 1.C2 = 0$ implies $C2 = 0$. Hence $C1 = 1$ and $C2 = 0$ is a solution, and thus $C1 \neq C2$ is again possible. ∎

Comment. This is a little surprising, because one would think that it is the nature of the primary equation that determines consistency/inconsistency and completeness/incompleteness. Problem: how to turn this into a condition on the null space of the primary matrix?

10.6 The Unit Equation

The results of the previous section suggest another test for inconsistency and incompleteness, one which is explanatorily more revealing. This test looks at the **unit equation** of (the secondary matrix of) the Necker. This is defined as the equation $M_s x = \mathbf{1}$, where x is a column vector of variables and $\mathbf{1}$ is the unit column vector having all ones.

Theorem 61 *A Necker is inconsistent iff its unit equation has solutions.*

Proof. Left to Right. If a Necker is inconsistent then either (a) some row of its primary matrix is all 1s, or the primary matrix is (b) the diagonal or (c) the antidiagonal matrix. If (a) then some column of its secondary matrix is all 1s. WLOG, let it be the first column, that for $C1$. Then $C1 = 1$ and $C2 = 0$ is a solution, since the equations are of the form $1.C1 + \text{any}.C2 = 1$. The argument is similar if it is the second column. If (b) then the equations are $1.C1 + 0.C2 = 1$ and $0.C1 + 1.C2 = 1$, which has $C1 = C2 = 1$ as a solution. Case (c) is similar. Right to Left. If a Necker is consistent then no row of M_p is all 1s, so no column of M_s is all 1s. Suppose that it has one 1 in it. If the other column is all zeros, then we have a situation like $1.C1 + 0.C2 = 1$ and $0.C1 + 0.C2 = 1$, which have no solutions. Similarly, if M_s is all zeros then the unit equation has no solutions, since each row would have to satisfy $0 + 0 = 1$. The final two cases are where the Necker is consistent and complete, but in this case the equations look like $1.C1 + 1.C2 = 1$ and $0.C1 + 0.C2 = 1$, which have no solutions. ∎

Theorem 62 *A Necker is complete iff the unit equation of its primary matrix has solutions.*

Proof. Left to Right. If a Necker is complete then no row of its primary matrix is all zeros. Thus M_p has 2 or more ones. If it has 4 ones then it is easy to see that either $R = 0, B = 1$ or $R = 1, B = 0$ are both solutions. If it has 3 ones then the row with the zero in it, *e.g.* $1.R + 0.B = 1$, forces $R = 1, B = 0$, which in turn satisfies the second row $1.R + 1.B = 1$. If it has two ones, then there are three cases: (a) two consistent and complete, (b) the diagonal, and (c) the antidiagonal. In case (a) we have e.g. $1.R + 0.B = 1$ and $1.R + 0.B = 1$ which is satisfied by $R = 1, B$ =any. In case (b) we have $1.R + 0.B = 1$ and $0.R + 1.B = 1$ which has the solution $R = B = 1$. Case (c), the antidiagonal, is the same. Right to Left: If a Necker is incomplete then at least one row of M_p is all zeros. This makes it impossible to satisfy the primary equation since for that row $0.R + 0.B = 1$, so there are no solutions. ∎

10.7 Conclusion

Theorem 63 supplies the desired assimilation of local and global inconsistency. The unit space of the secondary matrix is thus seen as the test of inconsistency of whatever kind. This justifies the use of the secondary equation, defined as representing ways of face building. To be inconsistent, both faces must have a value of 1 in the secondary equation. In contrast, to be consistent each face must have a value zero, corresponding to one face being consistently behind the other, or consistently in front. This prevents solutions of the unit equation. The contents of the unit space thus catalogue the ways that inconsistent Neckers are generated. This also generalises nicely to chains of Neckers, see Chapter 12. Additionally, Theorem 64 shows that the unit space of the primary matrix catalogues the ways that a Necker may be complete.

*

It is pleasing that 2×2 matrices over the simplest case of a field, namely Z_2, are adequate to describe the Escher cubes among the other Neckes. Indeed, the proofs in each case could be replaced simply by inspection of cases, but the proofs given above carry theoretical illuminations. In Chapter 12, we see what is preserved and what is lost in moving to the more general case of n Neckers. But first, we will look at yet another application of the Routley functor, to these matrices.

Figure 10.1: Stairs: three to five

Chapter 11

Linear Algebra & the Routley Functor

11.1 Introduction

We proceed to an analysis of the role of the Routley Star * on inconsistent and incomplete theories generated by the simplest case of a single Necker cube. It is shown here that the Routley Star is definable in terms of familiar linear-algebraic operations on matrices, and conversely, and that the Routley functor commutes with various combinations of these operations. These results constitute a further application of the Routley Functor outside its usual home, which contributes further to its explanation and justification as a mathematically useful tool in many areas.

In the previous chapter, it is shown how to represent the various consistency and completeness properties of Necker cubes by associating each Necker with a matrix, called the primary matrix M_p, over the field Z_2. Locally inconsistent Neckers were those with at least one row of M_p being a pair of ones $[1,1]$, incomplete Neckers were those with at least one row being a pair of zeros $[0,0]$, and the two globally inconsistent Neckers (the Escher cubes) were represented either the diagonal matrix Id that is $\begin{bmatrix} 1 & 0 \\ 0 & 1 \end{bmatrix}$ or by the anti-diagonal matrix $AntiId$ that is $\begin{bmatrix} 0 & 1 \\ 1 & 0 \end{bmatrix}$.

The former is standardly called Id in matrix algebra because it is the multiplicative identity, that is both left multiplication and right multiplication by Id leaves a matrix unchanged. We see the properties of AntiId as we go along.

11.2 Elementary Operations on Matrices

It will be recalled from linear algebra that on any matrix there are three *elementary row operations*, these being:

(1) **RowSwap** (swap a pair of rows)

(2) **RowAdd** (add one row to another)

(3) **ScalarMultiply** (multiply a row by a non-zero scalar)

It is noted that in Z_2, (3) is redundant, since the only non-zero scalar is 1. Furthermore, for 2×2 matrices, RowSwap and RowAdd take single matrices as arguments since there is only one pair of rows to be operated on. But RowAdd comes in two variants:

(4) **RowAdd1** (add one row to another and put the result in row 1

(5) **RowAdd2** (ditto but put the result in row 2).

Now we can see:

Theorem 63 *Let M be any 2×2 matrix over Z_2. Then:*

(a) $RowSwap(M) = AntiId \times M$

(b) $RowAdd1(M) = \begin{bmatrix} 1 & 1 \\ 0 & 1 \end{bmatrix} \times M$

(c) $RowAdd2(M) = \begin{bmatrix} 1 & 0 \\ 1 & 1 \end{bmatrix} \times M.$

Proof. These can be verified by straightforward matrix multiplication. ∎

Now we can identify further operations, particularly the column duals **ColSwap** (swap the columns), **ColAdd1** (add the columns and put the result in column 1) and **ColAdd2** (add the columns and put the result in column 2). It can now be proved that:

Theorem 64

(a) $ColSwap(M) = M \times AntiId$

(b) $ColAdd1(M) = M \times \begin{bmatrix} 1 & 0 \\ 1 & 1 \end{bmatrix}$

(c) $ColAdd2(M) = M \times \begin{bmatrix} 1 & 1 \\ 0 & 1 \end{bmatrix}$.

Proof. These may likewise be verified by straightforward matrix multiplication. ∎

One more matrix operation (not elementary) is **Switch**. This is the result of reversing all 1s and 0s in the matrix M. Think of Switch as the maximal switching act on an $n \times n$ array of switches.

Theorem 65 *Switch(M)=M+**1**, where **1** is the 2×2 matrix all of whose entries are 1s.*

Proof. Follows from the fact that in Z_2, 0+1=1 and 1+1=0. ∎

11.3 The Routley Star on Matrices

We return to the characterisation in Section 2 of a locally inconsistent Necker as having a primary matrix with at least one row being [1,1]. This arises because we identify matrices as being locally inconsistent if they have both colours R, B on at least one crossing, say $C1$. As noted in Chapter Nine, this is not inconsistent by itself, but it is inconsistent if one has as part of one's theory the *axiom of local consistency LoCon:* $(C)(R@C \rightarrow \neg B@C)$, along with its equivalent: $(C)(B@C \rightarrow \neg R@C)$. This means that the property of local consistency/inconsistency at the crossing Ci can be described by the equation: $a.R + b.\neg R = Ci$. Then we can form what can be called the *inconsistency equation* for a single Necker:

$$M_p \times \begin{bmatrix} R \\ \neg R \end{bmatrix} = \begin{bmatrix} C1 \\ C2 \end{bmatrix}$$

where M_p is its primary matrix. Clearly this is the primary equation with $\neg R$ substituted for B, but it has the merit of displaying explicitly the local inconsistency when Axiom LoCon is added.

If $a = b = 1$ we have the formal local inconsistency of having both R and $\neg R$ at Ci. This representation justifies in a more direct way the claim in Chapter Ten that local inconsistency in the principal matrix M_p and its associated Necker, comes to having a $[1, 1]$ row. Correspondingly, a row $[0, 0]$ indicates that the theory lacks both $R@C$ and $\neg R@C$ for the corresponding crossing C. That is, if $a = b = 0$, we have the formal incompleteness of the theory. This also justifies in a more direct way the claim in Chapter Ten that incompleteness in the primary matrix and its associated Necker comes to having a $[0, 0]$ row.

11.3. THE ROUTLEY STAR ON MATRICES

We recall that, if S is any set of sentences, then S^* is defined as $\{A : \neg A \notin S\}$. Now the star of an inconsistent theory is an incomplete theory, and vice versa. Reflected in the primary matrix, this motivates the definition:

Definition 66 *The Routley Functor * operating on matrices M, is the operation that switches all $[0,0]$ rows to $[1,1]$ rows, and vice versa, and leaves untouched all other rows.*

Now we have, (where dot . stands for functional application):

Theorem 67 $* = Switch.ColSwap = ColSwap.Switch.$

Proof. Consider the first equation. There are 4 cases, being four possible rows. Consider a row $[0,1]$. ColSwap produces $[1,0]$ and then Switch produces $[0,1]$, which is unchanged. The row $[1,0]$ is similar. Applied to the row $[1,1]$ ColSwap gives $[1,1]$ which when Switched gives $[0,0]$. Finally, ColSwap $[0,0] = [1,1]$ which when Switched gives $[0,0]$. Hence *= Switch.ColSwap. The second equation is a similar argument. ∎

That is, Switch commutes with ColSwap, and their product is the Routley Functor, which is thus definable in those terms. Also:

Theorem 68 $M^* = (M+1) \times AntiId = (M \times AntiId) + 1.$

Proof. From Theorem 67, adding 1 has the effect of Switch; the theorem then follows by applying Theorem 66. ∎

Theorem 69

(a) $Switch = *.ColSwap = ColSwap.*$

(b) $ColSwap = *.Switch = Switch.*.$

Proof. Part (a): It suffices that the equation holds for each row. So begin with any row of the form $[1,0]$. ColSwap produces $[0,1]$ which * leaves unchanged. The case $[0,1]$ is similar. ColSwap $[1,1] = [1,1]$ then applying * gives $[0,0]$. The case $[0.0]$ is similar. The proof of Part (b) is similar. ∎

Thus, from Theorems 69 and 71, the operators *, ColSwap and Switch form a pairwise commuting triplet such that each is the product of the other two. It is obvious that all operators are involutions. Hence we have that Id = Switch^2 = *^{2} = ColSwap^2 = Switch.*.ColSwap etc. The three operators thus enjoy a close relationship, which further vindicates the Routley * as a natural operation on these matrices. In the next chapter, it will be seen that this relationship is continued when moving to larger figures and matrices.

Definition 70 *Anti* is the operation that does to columns what Routley * does to rows, that is Anti* switches [0,0] columns to [1,1] columns and vice versa.*

Theorem 71

(a) $Switch.RowSwap = RowSwap.Switch = Anti^*$

(b) $Switch = Anti^*.RowSwap = RowSwap.Anti^*$

(c) $RowSwap = Anti^*.Switch = Switch.Anti^*$

(d) $Id = Switch^2 = Anti^{*2} = RowSwap^2 = Switch.Anti^*.RowSwap$
etc.

Proof. These can all be verified by matrix calculation, or by symmetry from the previous theorem. For example, part (a): It suffices to prove the equation for each column. So begin with any column of the form of the column vector col[1,1]. Applying RowSwap gives col[1,1] and then applying Switch gives col[0,0] which is Anti*(col[1,1]). The argument for col[0,0] is similar. Next take col[1,0]. Applying RowSwap gives col[0,1] then applying Switch gives col[1,0] which we began with. But this is the same as Anti*, which leaves such rows unchanged. The argument for col[0,1] is similar. ∎

This means that the Routley Functor has a natural dual in the dual space, which is definable in terms of Switch and the elementary row operation RowSwap. Again we see an example of three operations closely related. The previous results can now be combined, in virtue of the fact that Switch is common to both spaces, to give results like:

Theorem 72

(a) $*.ColSwap = Anti^*.RowSwap$

(b) $* = Switch.*.Switch$

11.4 Transposes

We recall from linear algebra the well-known matrix operator Transpose, that is T, defined as the result of swapping rows for columns across the main diagonal; that is, element a_{ij} becomes element a_{ji}. Transpose is clearly an involution, that is $M^{TT} = M$. We can also define **Antitranspose**, shortened to $AntiT$, as the operation which swaps rows and columns across the antidiagonal. Now we have:

Theorem 73

(a) $RowSwap.T=T.ColSwap$, and similarly with $AntiT$ for T.

(b) $RowSwap=T.ColSwap.T$, and similarly for $AntiT$

(c) $T=RowSwap.T.ColSwap$, and similarly for $AntiT$

(d) $Anti*.T=T.*$

(e) $Anti*=T.*.T$

(f) $Switch.T=T.Switch$

(g) $AntiId=Switch(Id)=RowSwap(Id)=ColSwap(Id)$.

(h) $AntiId^2=Id$.

Proof. For (a), we have:

From $\begin{bmatrix} a & b \\ c & d \end{bmatrix}$ first applying ColSwap gives $\begin{bmatrix} b & a \\ d & c \end{bmatrix}$ then applying T gives $\begin{bmatrix} b & d \\ a & c \end{bmatrix}$ whereas first applying T gives $\begin{bmatrix} a & c \\ b & d \end{bmatrix}$ then applying RowSwap gives the same $\begin{bmatrix} b & d \\ a & c \end{bmatrix}$

Part (b) follows by applying T on the right to both sides of (a). Part (c) follows by applying ColSwap on the right to both sides of (a). For Part (d), it is clear that first swapping [1,1] rows for [0,0] rows then transposing, is the same as first transposing, then swapping [1,1] columns for [0,0] columns. Part (e) follows from (d) by application of T on the right. Parts (f) and (g) are obvious. Part (h) may be verified by matrix multiplication. ∎

Comment. Note that in general $T.* \neq *.T$, that is T and Star do not commute; this can be seen by applying these successively to $\begin{bmatrix} 1 & 1 \\ 0 & 0 \end{bmatrix}$.

11.5 Order, Covariance and Contravariance

There is a natural order on matrices:

Definition 74 *For matrices M, N we can define $M \leq N$ to mean that each element of M is $\leq$ the corresponding element of N, and $M < N$ to mean $M \leq N$ and $M \neq N$. An operation is said to be **covariant** if it is order-preserving, and **contravariant** if it is order-reversing.*

The question then arises: which of the above operations are covariant and which contravariant?

Theorem 75 $M < NQ$ *iff* $Switch(N) < Switch(M)$ *(Switch is contravariant).*

Proof. If $M < N$ then M has 0s in some places where N has 1s, and they otherwise agree. Hence $Switch(N)$ has 0s in some places where $Switch(M)$ has 1s and they otherwise agree. Hence $Switch(N) < Switch(M)$. ∎

Theorem 76 *The following are covariant: RowSwap, ColSwap, T, AntiT.*

Proof. It is evident that these operations preserve the order as they preserve which entry corresponds to which. ∎

Theorem 77 *The Routley * and Anti* are contravariant.*

Proof. That the Routley $*$ is contravariant follows from Theorem 69, namely that $* = $ Switch.ColSwap. That is, ColSwap is order-preserving, and then Switch reverses the order. That Anti* is contravariant is a similar argument. Alternatively, we may invoke Theorem 75(e), namely Anti$* = T.*.T$, for then T preserves order, $*$ reverses it, and finally T again preserves it. ∎

11.6 Conclusion

It is an appropriate result that the Routley Functor on matrices is contravariant, since its counterpart on theories is order-reversing under the subset relation $\subset$ as order. But the derivation of this fact from the contravariance of Switch together with the covariance of other matrix

11.6. CONCLUSION

operations is novel, as is the dual result for Anti*. These results demonstrate the applicability and richness of the Routley Functor interacting with linear algebra, which adds to the motivation that we have seen for it up to now.

*

In the next chapter, these methods are extended to the representation of arbitrary collections of Neckers, and the special case of several Neckers chained together. It is seen that some modification is necessary, but also that some of the representations of the Routley Functor remain invariant.

Figure 11.1: Dimensions

Chapter 12

Necker Chains

12.1 Introduction

In the two previous chapters, the Escher cubes and their Necker variants were studied from the point of view of linear algebra over Z_2. In this chapter, these methods are extended to the general case of n Neckers. We begin with an obvious point, that the most general case of n Neckers, coloured differently and distributed randomly across a page, does not have much interest beyond what has already been shown: they are simply described by n matrices, each 2×2. These could be arranged artificially into a single $2n \times 2n$ matrix over Z_2, but its 2×2 cells would have no interaction with one another. The possible diagrams that can be drawn with n Neckers are too diverse to get much theoretical purchase on, beyond the case of the single Necker.

However, there is a more interesting special case, still a generalisation of Escher cubes, that displays an additional higher order inconsistency. This higher order inconsistency is itself a generalisation of the global inconsistency of the Escher cubes described in Chapter Nine, which arises not at a single crossing, but from the interaction between two crossings, each consistent in itself. The special case we study is the case where n Neckers of similar colours are chained together in a row. We can call this an **n-chain**, and each Necker from which it is constructed is called a **cell**. It should be stressed that the colouring of the Neckers is not essential. As was seen in Chapter Nine, it serves only as a convenient vehicle for disambiguation of the individual Neckers, that is the case $n = 1$, by keeping track of faces, which can in any case be achieved by the device of breaking the occluded edge as in knot theory. Such higher order inconsistency is manifested in the case $n = 2$, as in Figure 12.1, where each individual Necker cell is entirely consistent, only the relation between the two is impossible.

12.2. CHAINED NECKERS

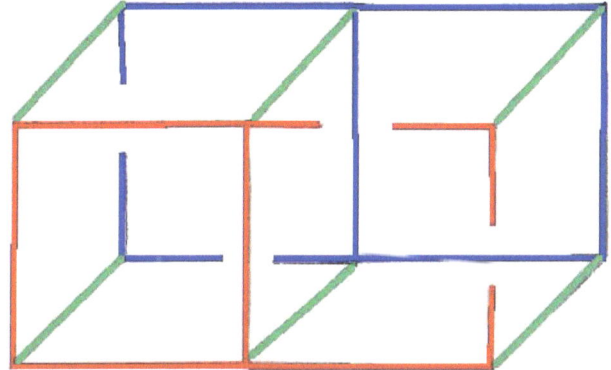

Figure 12.1: Higher order inconsistency in a chain

We use this $n = 2$ case as an illustration of the results that follow.

12.2 Chained Neckers

The two-chain of Neckers in the above diagram has four crossings. As in Chapter 9, each crossing can have both colours (local inconsistency), neither (local incompleteness), or one (local consistency and completeness). It is clear that the whole pattern of crossings can be described in the following matrix equation:

$$\begin{bmatrix} a & b \\ c & d \\ e & f \\ g & h \end{bmatrix} \times \begin{bmatrix} R \\ B \end{bmatrix} = \begin{bmatrix} C1 \\ C2 \\ C3 \\ C4 \end{bmatrix}$$

The left hand 4×2 matrix can be called the *primary matrix* M_p of the diagram. It consists of two 2×2 cells arranged vertically, each being the primary matrix of the corresponding single Necker cell of the diagram. The same information can be displayed in four individual equations:

a. $R + b.B = C1$

c. $R + d.B = C2$

e. $R + f.B = C3$

g. $R + h.B = C4$.

It is noted that for the general case of n Neckers chained horizontally, there are $2n$ crossings and a primary matrix of dimension $2n \times 2$. All of the coefficients $a...h$ come from Z_2; this is all that is necessary since each crossing is formed from the contribution of just two colours.

In moving to the secondary matrix, the appropriate generalisation takes into account the contributions of red and blue crossings. We recall that in the $n = 1$ case, the secondary matrix M_s was the transpose of the primary matrix, but that is not available here because the primary matrix is not square. However, we can proceed via an intermediate construction. Each of the two cells, taken separately, makes up a red face RF and a blue face BF, so there are 4 faces, $RF1$ and $BF1$, and $RF2$ and $BF2$. In the $n = 1$ case, that is the 2×2 construction of Chapter 10, each cell is described by its secondary equation $M_p^T \times C = F$, where C is the 2×1 column vector of crossings, and F is the 2×1 column vector of faces. Thus in the $n = 2$ case we have both: $M1_p^T \times \begin{bmatrix} C1 \\ C2 \end{bmatrix} = \begin{bmatrix} RF1 \\ BF1 \end{bmatrix}$ and $M2_p^T \times \begin{bmatrix} C3 \\ C4 \end{bmatrix} = \begin{bmatrix} RF2 \\ BF2 \end{bmatrix}$.

But now we postulate that the large red face, as the composite of the red faces, is given by the sums of their crossings, and hence the sums of their faces: $RF = RF1 + RF2$. Similarly the large blue face is the sum of individual blue faces: $BF = BF1 + BF2$. It is clear how to write this in a matrix equation; it is simply:

$$[M1_p^T, M2_p^T] \times C = F$$

The (left hand) matrix is the secondary matrix M_s having dimensions 2×4, C is the 4×1 column vector of the four crossings and F is the 2×1 column vector of large faces. This is the *secondary equation* for the chain of Neckers. In the case of n Neckers, under the assumption that faces add linearly, the secondary matrix is $2 \times 2n$, C is $2n \times 1$, and F remains the same.

The results of Chapter 10 indicate what is important for inconsistency, either local or higher order (including global). The *unit equation* is the secondary equation $M_s \times C = F$ where F is the unit vector $\mathbf{1}$ of all ones. We want to show that a Necker chain is inconsistent iff its unit equation has solutions for C; and equivalently, a chain is consistent iff its unit equation has no solutions for C.

To get to this, first we need a sense of which Necker chains are consistent and which are inconsistent. Every consistent Necker chain must have all its component cells consistent, since if any cell is inconsistent then the whole has that as an inconsistent part. (The reverse fails, as we can see from Figure 12.1, which has each of its component cells consistent: they are simply chained together in an inconsistent way, in which

12.2. CHAINED NECKERS

the blue face dominates in one case but the red face dominates in the other.) There are two basic types of consistent Necker chains. (1) There are the two basic consistent Neckers, one in which all red faces and edges are unambiguously in front of the blue faces and edges where they occlude, and the other in which it is blue that dominates. (2) In addition, there are all those which can be obtained from (1) by the deletion of one or more colours at a crossing. These are incomplete, since they have crossings at which there are no colours; but they remain consistent, since they are deletions from cases which are already consistent. See Figure 12.2 for two consistent chains, the left hand chain is complete, the other has its rightmost cell incomplete.

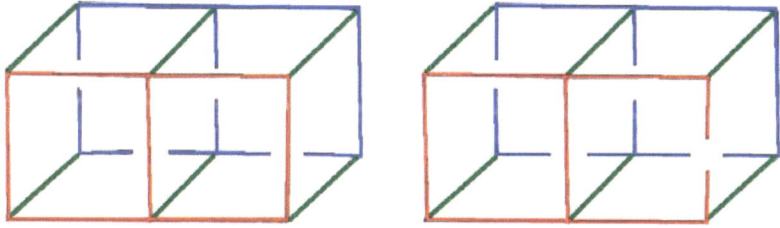

Figure 12.2: Two consistent Necker chains

All the Necker chains other than types (1) and (2) are inconsistent, and they come in two further forms (3) Red and blue edges both appear at the same crossing. These are inconsistent in virtue of having an inconsistent part, namely that individual crossing. Alternatively, (4) Red and blue never appear at the same crossing, but red is unambiguously in front at one crossing, and blue is unambiguously in front at another crossing. This is inconsistent because no large face can be put together which has the same colour unambiguously in front along the whole chain. The figure we began with is an example. It is clear that these four possibilities are exclusive and exhaustive.

Now we need a lemma.

Lemma 78 *The unit equation has no solutions iff one or both of the rows of M_s is all zeros.*

Proof. R to L: if one of the rows of M_s is all zeros, there is no way to assign numbers to the column vector C of crossings to produce a 1 in that row of F. Hence the secondary equation has no solutions. L to R: Conversely, suppose that neither row is all zeros. So each row has at least one 1 in it. To find a solution to the secondary equation,

(a) find the first column from the left that is both 1s, and put a 1 in that row of C, then complete C with zeros. It is clear that this C corresponds to the pair of equations $0 + 0 + ... + 1.1 + 0 + ... + 0 = 1$ for the first row, and identically for the second row, which is a solution. Otherwise (b), find the first 1 in the first row, say column i, then the first 1 in the second row, which must be in a different column j from that in the first row. Put 1s in each of the corresponding rows i, j of C, and complete C with zeros. This corresponds to the pair of equations $0 + 0 + ... + 1.1(ith.place) + 0 + ... + 0 = 1$, and $0 + 0 + ... + 1.1(jth\ place) + 0 + ... + 0 = 1$. Again this is a solution. ∎

Now we have the main theorem of this section.

Theorem 79 *A chain of Neckers is inconsistent iff its unit equation has solutions.*

Proof. L to R: From our observations on consistent chains above, if a chain has a secondary matrix with a row of zeros, then it is consistent. So if it is inconsistent, then no row is all zeros. Hence by the lemma its unit equation has solutions. R to L: Conversely, if the unit equation has solutions, then by the lemma no row of M_s is all zeros. But we also noted above that if a secondary matrix has no row all zeros, then it is inconsistent. ∎

12.3 Degree of Inconsistency

The previous result motivates the following definition.

Definition 80 *The **degree of inconsistency** of a Necker chain is the number of independent solutions of its unit equation.*

This seems intuitively reasonable: the unit equation tracks the independent ways in which both of the large faces get a non-zero value in the secondary equation. If the product of the secondary matrix with the column vector of crossings inevitably yields a zero face no matter what the state of the crossings, then it must have a row of zeros and so be consistent.

The test of the definition is in the results it produces, so we compute the degree of inconsistency for several salient cases.

(i) For individual Neckers, that is cells, we have that the trivial Necker has as its secondary matrix $\begin{bmatrix} 1 & 1 \\ 1 & 1 \end{bmatrix}$, which has solutions for $C = \begin{bmatrix} 1 \\ 0 \end{bmatrix}$ and

12.3. DEGREE OF INCONSISTENCY

$\begin{bmatrix} 0 \\ 1 \end{bmatrix}$. These are evidently independent of one another, so the number of independent solutions is at least 2. It is at most 2 also, since there are no more independent solutions for 2×1 column vectors. Hence it is exactly 2. This is the maximum degree of inconsistency for figures of this size, which accords with the intuition that the figure with red and blue at all crossings is maximally inconsistent.

(ii) For the important case of Escher's cube, that is the globally inconsistent Necker whose secondary matrix is Id, (Figure 12.3):

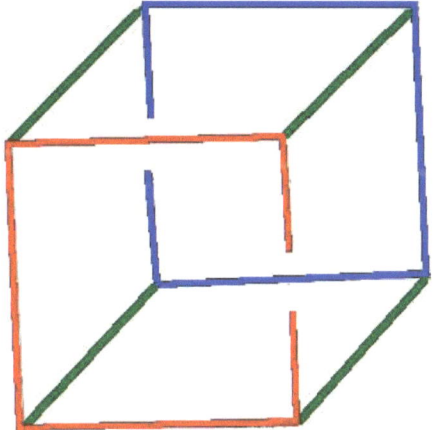

Figure 12.3: Escher's cube

Its secondary matrix is $\begin{bmatrix} 1 & 0 \\ 0 & 1 \end{bmatrix}$, and we can compute that there is just one solution, $C = \begin{bmatrix} 1 \\ 1 \end{bmatrix}$. Thus the degree of inconsistency is 1, the intermediate value.

(iii) For the consistent Neckers having at least one row of zeros, it is impossible to get solutions for the unit equation, so degree of inconsistency $= 0$.

(iv) The case $n = 2$ (maximum independent solutions $= 4$) contains more possibilities, and illustrates higher order inconsistency. For example, there is the trivial Necker chain (Figure 12.4).

This has solutions $C = $ any of: $\begin{bmatrix} 1 \\ 0 \\ 0 \\ 0 \end{bmatrix}, \begin{bmatrix} 0 \\ 1 \\ 0 \\ 0 \end{bmatrix}, \begin{bmatrix} 0 \\ 0 \\ 1 \\ 0 \end{bmatrix}$ and $\begin{bmatrix} 0 \\ 0 \\ 0 \\ 1 \end{bmatrix}$.

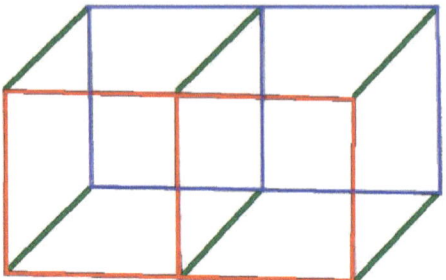

Figure 12.4: The trivial Necker 2-chain

These solutions are all independent, so degree of inconsistency = 4, which is maximal as triviality should be.

(v) Any chain with the same colour failing to appear at all crossings (ie. consistent), has a row of zeros and so has a degree of inconsistency = 0, which is appropriate.

(vi) Any chain with both colours at a single crossing C_i has a solution with the value for row $i = 1$ and zeros elsewhere. Hence, with both colours at say 3 crossings, it has 3 indendent solutions with a single 1 in that place, and has a degree of inconsistency = 3. That is, adding inconsistent crossings increases the degree of inconsistency.

(vii) Another important case is Figure 12.1 we began with, which displays higher order inconsistency. The secondary matrix for this figure is: $\begin{bmatrix} 1 & 1 & 0 & 0 \\ 0 & 0 & 1 & 1 \end{bmatrix}$. Its four solutions are: $\begin{bmatrix} 1 \\ 0 \\ 0 \\ 1 \end{bmatrix}$, $\begin{bmatrix} 0 \\ 1 \\ 1 \\ 0 \end{bmatrix}$, $\begin{bmatrix} 1 \\ 0 \\ 1 \\ 0 \end{bmatrix}$ and $\begin{bmatrix} 0 \\ 1 \\ 0 \\ 1 \end{bmatrix}$.

However, each of these is the sum of the others, so any one can be eliminated. Hence, degree of inconsistency = 3. Compare with Id above.

Conjecture: any n-chain whose cells are copies of Id, (The Escher cube) has a degree of inconsistency n.

12.4 Elementary Row and Column Operations

We return to the properties of the Routley Functor and related operators. The results of the previous chapter cannot all remain true, because the primary and secondary matrices are not square, so that, for example, Transpose is not defined. Moreover, since matrices vary in size, operators must be defined differently for different sizes.

12.4. ELEMENTARY ROW AND COLUMN OPERATIONS

We begin with the three elementary row operations on the space of row vectors. First, **RowSwap$_{ij}$(M)** is the result of swapping rows i and j in matrix M. We suppose that M is $m \times n$. What matters for RowSwap is the number of rows m on the matrix whose rows are being swapped. The *basic* swap operation is performed on Id_m, which is the $m \times m$ square matrix with 1 down the main diagonal and zeros elsewhere. $RowSwap_{ij}(Id_m)$ is thus the result of deleting the 1s at places (i,i) and (j,j) and replacing them by 1s at (i,j) and (j,i). It is then a simple matter to verify by matrix multiplication that $RowSwap_{ij}(M)$ can be achieved by left multiplication of M by $RowSwap_{ij}(Id_m)$.

The second elementary row operation is **RowAdd1$_{ij}$**, which is the result of adding row i to row j, and putting the result in row i. To define this, take Id_m and add a 1 in place $(i,j) = Id_m + [a_{ij} = 1]$. Then it is a simple matter to verify by matrix multiplication that left multiplication of M by this matrix has the effect. The matrix is also the result of applying RowAdd to Id itself. In Chapter 10, there is also distinguished **RowAdd2$_{ij}$**, which adds row i to row j and puts the result in row j. But since RowAdd2$_{ij}$= RowAdd1$_{ji}$, this is not proceeded with.

The third elementary matrix operation is scalar multiplication. But since we remain in linear algebra over Z_2, then as in the single-cell Necker, the only nonzero scalar is 1, so this operation is redundant.

These can be summarised as:

Theorem 81

(a) $RowSwap_{ij}(M) = RowSwap_{ij}(Id) \times M$

(b) $RowAdd_{ij}(M) = RowAdd_{ij}(Id) \times M = (Id + [a_{ij} = 1]) \times M$.

There are also elementary column operations **ColSwap** and **ColAdd**. They can also be defined in terms of (right) multiplication by a suitable matrix. In each case, the effect is obtained by performing the same operation on Id_n, then right multiplication of M by this. Note the index n of Id_n. This is necessary as right multiplication of a matrix M which is $m \times n$, requires that Id have n rows.

Then there is the useful and natural operation **Switch**, (not an elementary operation). This is the result of changing all 0s to 1s and versa. This effect can be obtained by adding $\mathbf{1}_{mn}$, the $m \times n$ matrix of 1s. In summary:

Theorem 82

(a) $ColSwap(M) = M \times ColSwap(Id)$

(b) $ColAdd(M) = M \times ColAdd(Id)$.

(c) $Switch(M) = M + 1$

12.5 The Routley Functor

We recall that the Routley Functor interchanges rows (1,1) for (0,0), leaving rows (0,1) and (1,0) alone. Clearly then, in the general case, the Routley functor operates naturally only on matrices with 2 columns. Thus it operates on the primary matrix M_p, which is $2n \times 2$.

In passing, we note that the effect of applying this operation to a chain of Neckers is to produce another chain, the chain obtained taking all crossings lacking any colour, and filling in with both colours, and taking any crossing with both colours and erasing both colours. That is, the Routley star is a natural transformation from Necker chains to Necker chains.

But it also has natural relationships with the above linear algebraic operators.

Theorem 83

(a) * = Switch.ColSwap = ColSwap.Switch.

(b) Switch = *.ColSwap = ColSwap.*

(c) ColSwap = *.Switch = Switch.*

(d) * = Switch.*.Switch

Proof. On (a): It is clear that operating on rows (1,1) and (0,0) in either order changes either into the other. On the other hand, operating on rows (0,1) with ColSwap changes it to (1,0) and then Switch changes it back. Similarly for the opposite order of operation. Similarly for operating on rows (1,0) and (0,1). On (b): ColSwap(1,1) = (1,1), and then (1,1)* = (0,0) which is the effect of Switch on the original. The other rows are similar. On (c): Switch(1,0) = (0,1), and then (0,1)* = (0,1) which is the effect of ColSwap on the original. The other rows are similar. On (d): * has no effect on rows (0,1) and (1,0), so the effect is two consecutive applications of Switch, which is the identity. But for rows (1,1) and (0,0), * toggles between them, as does Switch. ∎

We recall that there is also Anti*, which does to columns what * does to rows. This means that Anti* only operates on matrices with two rows. With appropriate restrictions on the size of matrices then, we have, using similar calculations to the previous theorem:

Theorem 84

(a) $Anti^* = Switch.RowSwap = RowSwap.Switch.$

(b) $Switch = Anti^*.RowSwap = RowSwap.Anti^*$

(c) $RowSwap = Anti^*.Switch = Switch.Anti^*$

(d) $Anti^* = Switch.Anti^*.Switch$

Note that we cannot put together part (b) of the two previous Theorems to draw the conclusion as in Chapter 11 that *.ColSwap = Anti*.RowSwap, because the size of the Switch matrix is in general different on the LHS from the RHS.

However, we can say that the above results linking the Routley functor with other linear algebra operators remain true as generalisations of the results of Chapter 11, and thus demonstrate the naturalness in this setting of the Routley functor and its dual, Anti*.

12.6 Order, Covariance and Contravariance

Matrices of the same dimensions have the same definition of order as in Chapter 11, and the same arguments given there continue to hold. So we have:

Theorem 85

(a) The following operators are covariant: RowSwap, ColSwap

(b) The following operators are contravariant: *, Anti*, Switch.

12.7 Conclusion

The main results of this chapter are the description of chains of Neckers, the identification of higher-order inconsistencies arising from the impossible juxtaposition of consistent Necker cells, the definition of degree of inconsistency, and the method of calculation of degree of inconsistency as illustrated by a number of salient cases. In addition, it was seen that the Routley functor operates naturally on Necker chains, and is interdefinable in this context with other linear algebra operators, thus demonstrating further its broad usefulness in the theory of inconsistency.

*

This completes our study of the Necker cubes. In the next chapter, we discuss a different kind of impossible picture, namely the triangle. We see that it is a kind of generalisation of the Necker, but a dimension higher.

Figure 12.5: Chains and tesselation

Chapter 13

The Triangle

13.1 Introduction

Figure 13.1: The triangle

In this chapter we deal with The Triangle. It will first be proved that The Triangle is a genuine paradox. To do that, it is necessary to provide a logical analysis according to which the figure is correctly described by some proposition A and its denial $\neg A$. It will be recalled that this was also the first task in the analysis of global inconsistency in Necker cubes, back in Chapter 9. Having done this, we go further and prove that the triangle is an occlusion paradox, that is one which can be rendered non-paradoxical by the modification of one or more occlusions. Then it is seen how this logical analysis gains support from a series of findings in cognitive science. This indirectly confirms the methodological assumption noted in Chapter 8, that inconsistent images derive their inconsistent nature from the contents of cognitive states. Finally, it is necessary to progress beyond logic, by providing a mathematical theory of the triangle, from which the logical analysis follows. We round off the chapter by discussing some of the literature on the triangle, with an eye to seeing what not to say about the triangle.

13.2 The Triangle is a Paradox

There are a number of intuitions about the above image (Figure 13.1) that suggest a contradiction. However, one key intuition is that it constantly recedes from the observer as the eye traverses the image in an anticlockwise direction. For a closed loop to do this is impossible. This intuition can be made more precise in the following manner.

To show that the triangle is a genuine paradox, that is it is described by some A and $\neg A$, it is necessary to label some of its parts (Fig 13.2). We identify three points A, B, C on the corners of the diagram. Points A and B lie on face (plane) AB, B and C lie on face BC, and C and A lie on face CA. Axes x and y are the usual axes in the plane of the paper, and the z-axis is normal to the plane of the paper.

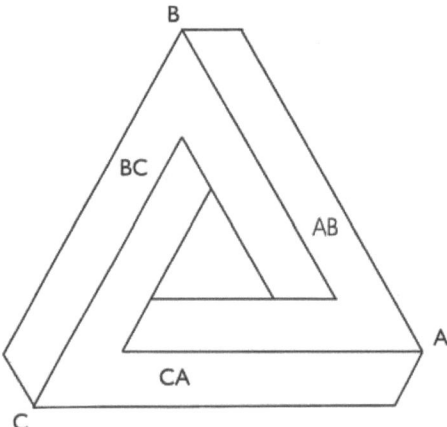

Figure 13.2: The triangle labelled

We noted in Chapter 8 that Cowan and Pringle (1978) identify several types of corner. Corners are key in this chapter too. The corner A, considered in 3-D, tilts **into** the page, when the corner A is traversed in the anti-clockwise direction. This means that points further along the face AB are futher away in the z-direction than points not so far along. Thus we can deduce that:

(1) A is closer in the z-direction than B.

13.3. THE TRIANGLE IS AN OCCLUSION PARADOX

Traversing the corner B anti-clockwise, the face BC similarly tilts further into the page. Hence:

(2) B is closer in the z-direction than C.

And traversing the corner C anticlockwise, the face CA similarly tilts further into the page. Hence:

(3) C is closer in the z-direction than A.

We now make the natural assumption, that the relation "closer" is transitive:

(4) Trans (closer).

We therefore deduce:

(5) A is closer in the z-direction than A.

But evidently, by observation (or by definition of *closer*):

(6) A is not closer (in the z-direction) than A.

The statements (5) and (6) contradict one another. This is the proposed analysis of the paradoxicality of the triangle.

13.3 The Triangle is an Occlusion Paradox

Now we prove that it is an occlusion paradox. By definition, this means that it can be rendered non-paradoxical by changing one of more occlusions. In fact, the present case requires two changes. in the course of the proof we identify an intermediate case of another triangle that is also paradoxical. Both are seen to be occlusion paradoxes. The argument proceeds in two stages. Consider first the bottom left hand corner C (Figure 13.3).

Figure 13.3: Bottom left hand corner

We can change the tilt of this corner from *into the page moving anticlockwise*, to *into the page moving clockwise*. See Figure 13.4.

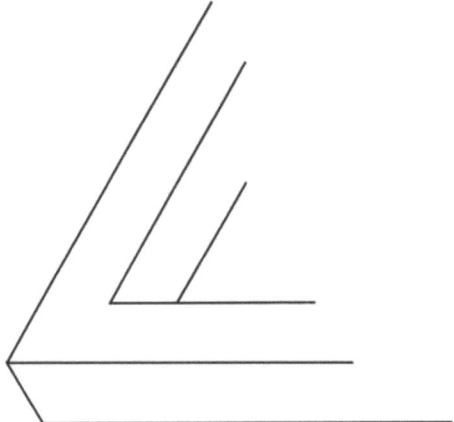

Figure 13.4: Modified corner element

Replacing the old corner C by this new corner gives a new triangle. See Figure 13.5.

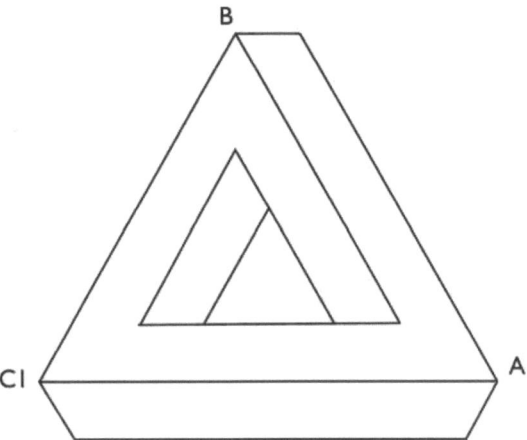

Figure 13.5: Modified triangle, inconsistent

Another Paradoxical Triangle

This triangle is also paradoxical. However, to show this requires some modification to the proof. The argument given for the paradoxicality

13.3. THE TRIANGLE IS AN OCCLUSION PARADOX

of the first of the two triangles no longer works. This is because the change to the C-corner renders unsupported the premiss that the point C is closer in the z-direction than the point A. But a different argument works. For the z-axis, choose a normal to the A-B face. This allows us to say:

(1) A is the same distance in the z-direction as B

Now the B corner is still turned in, when traversed in an anti-clockwise direction. So:

(2) B is closer in the z-direction than $C1$.

But now, inspection makes it clear that $C1$ is part of the same plane as the AB face, therefore with the same normal. Hence:

(3) $C1$ is the same distance in the z-direction as A.

Now we invoke a weaker premiss than transitivity, a form of *functionality*, namely that *if A is the same distance as B and B is closer than $C1$ then A is closer than $C1$*. Hence:

(4) A is closer in the z-direction than $C1$.

But this is incompatible with (3).

Both Triangles are Occlusion Paradoxes

Returning to the proof that the paradoxes are occlusion paradoxes, we note that the above change of the C-corner is affected by changing some occlusions. Two edges which were unoccluded become occluded, and two edges which were occluded become unoccluded. This is indicated by arrows on the diagram (Figure 13.6).

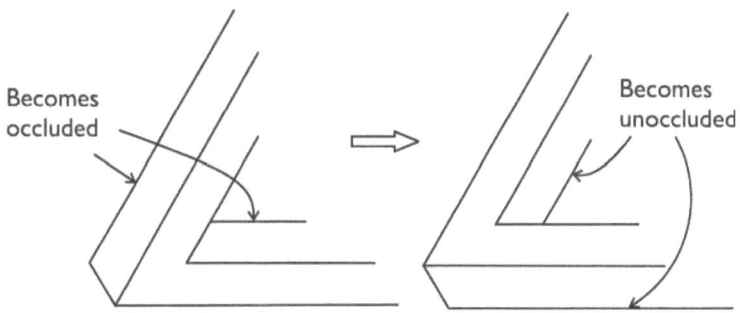

Figure 13.6: First occlusion changes

In passing, the changes to the occlusions of the edges carry with them changes to occlusions of the faces bounded by those edges. This suggests a further analysis in terms of occluded faces which we will not pursue here.

To complete our argument requires a change of the occlusions at the top corner B, opening it up (Figure 13.7).

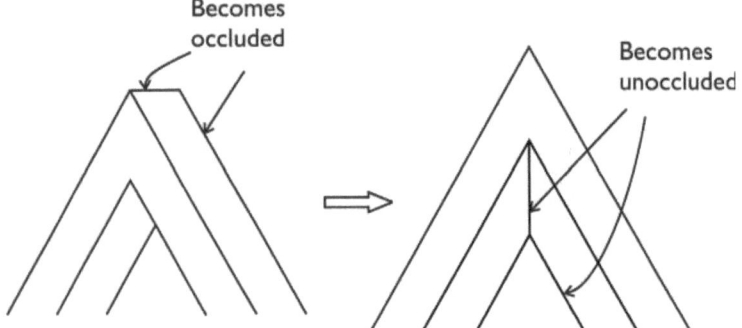

Figure 13.7: Second occlusion changes

It is clear that the final image (Figure 13.8) is consistent.

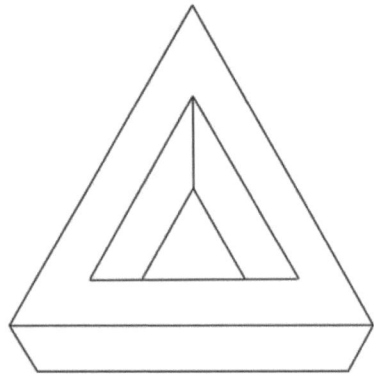

Figure 13.8: The final outcome: consistent

Summing up this argument:

Theorem 86 *The two paradoxical triangles are occlusion paradoxes.*

It is clear that a similar analysis of occlusion paradoxes of 4 sides of more is available, and that that we can make similar changes of occlusions to such figures to render them consistent or inconsistent, as the case may be.

13.4 From Cognitive Science to Logic

My claim has been that for impossible images the logical analysis reflects, and is justified by, the state of cognition. Cowan and Pringle (1978) provide just the sort of experimental data relevant to this claim. They work with 4-sided figures, but 3-sided figures are obviously a special case. They identify configurations made up of corner-bar-corner, traversed in a given fixed direction, such as anticlockwise. These can be seen to be turning into the plane of the paper, out of it, or coplanar. They define the *parity* of such a configuration, to be $+1$, -1 or zero according as it turns into, out of, or coplanar with the plane of the paper. The parity of a figure is then the sum of the parities of its configurations.

The question they ask is: is the parity of a figure related to the judgement of degree of *possibility* (on a scale of 0-10) of the figure? Twenty-seven different 4-sided figures (Figure 13.9) were the set over which degrees of possibility were judged, with 3 groups of judges. Each judge was assigned to a subset of 18 of these, and then the other 9 were used to calibrate the judges, as instruction providing standards of "possibility" (consistency), and degrees of inconsistency.

It is conceivable that the judges could have rejected the training standards, as for any given figure, a majority of judges were not trained on it. But this was not so. In fact, the result was a strong inverse correlation between parity and possibility. That is, the greater the parity (that is the more that corners turn into or out the plane of the page), the lower the judgement of possibility. Figure 13.10 shows the systematic relation between judged degree of possibility/impossibility and the above ordering of the 27 images.

We should also take note of the presence in this experiment, of judgements of *degrees* of possibility and impossibility. Classical logicians, holding as they do that the inconsistent has no structure, must say that this is a confusion. But it is no error. For instance the full impossible triangle above (Figure 13.1) has a parity of 3, in contrast to which the halfway corrected triangle (Figure 13.5) has a parity of 1: as we noted the latter triangle is still impossible (which is why a further change of occlusions was necessary), but its impossibility is intuitively of a lesser kind.

The correlation between impossibility and parity was compared with two rival explanations of varying judgements of possibility. One explanation was that possibility is related to *contortion number* of the figure. Contortion is the number of twists in the sides necessary for the corners to appear as they do while the whole figure remains coplanar. The difference between contortion and parity, is that opposite parities in a figure cancel out, whereas contortions accumulate (Figure 13.10 shows both numbers, bottom). However, when contortion was compared with judgements of possibility, there was no correlation.

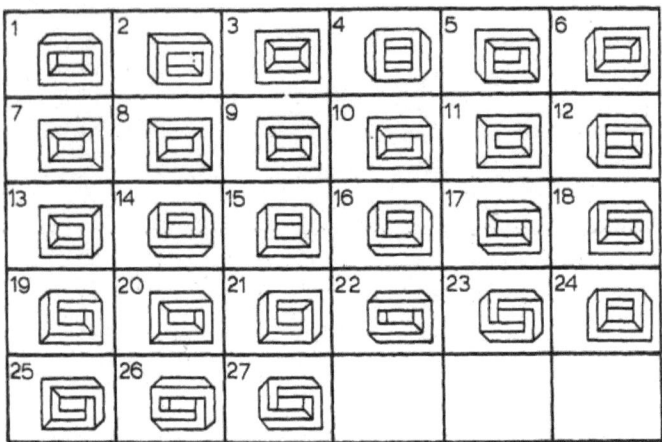

Figure 13.9: Cowan and Pringle's 27 rectangles

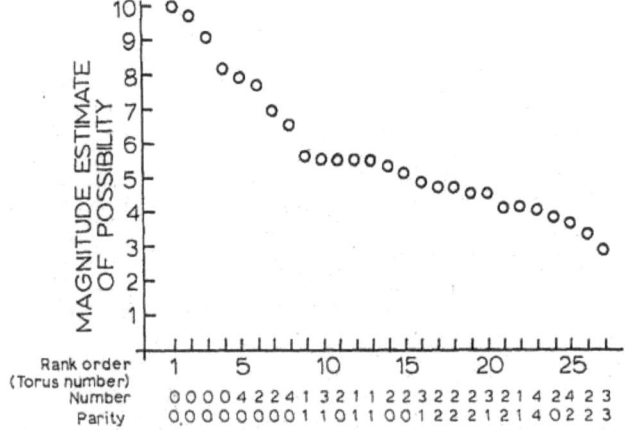

Figure 13.10: Cowan and Pringle's results

Another explanation raised was that judgements of possibility are affected by the presence or absence of stereopsis, which can be induced by the usual stereographic methods. But not only was there no significant correlation with stereopsis, but the same correlation between possibility and parity was found in the presence of stereopsis. This thus reinforces the earlier conclusion.

In short, it is the cognition of parity that explains judgements of possibility and impossibility. The latter judgements are unaffected by contortion and by stereopsis. But this is just what is wanted. The logical analysis in the first two sections of the chapter takes advantage of it. If a configuration has a parity of 1, then it turns in, and then the logical analysis proceeds as indicated.

13.5 From Logic to Mathematics

We have seen a logical analysis of the paradox we began with, and evidence from cognitive science that this analysis is correct in terms of the cognitive perceptual states generated by the paradoxical image of the triangle. To complete the story, it is necessary to see how an inconsistent mathematical theory can explain the logic-cognition nexus. This provides evidence for our leading thesis, that we internalise an inconsistent mathematical theory when perceiving inconsistent images. We can use a suggestion of George Francis (1987, 68-9). This suggestion has already been encountered in Chapter 8, and it is developed further here.

To fit with Figure 13.11 (adapted from Francis), it is convenient to think of the canonical triangle rotated clockwise through 60°, so that it balances on the corner A. Francis asks the question: in what sort of (consistent) space could the triangle occur? He answers that it could occur in $R^2 \times S^1$. This is space rolled up as a cylinder in the z-direction. If we unroll the S^1 dimension, we get an infinite repetition, which is represented in the diagram.

Setting the z-axis to be the dotted line through $A, D, D', \ldots$ we can say:

(1) A is closer in the z-direction than B

(2) B is closer in the z-direction than C

(3a) C is closer in the z-direction than D.

Now roll the z-direction back up into S^1. This amounts to identification of A with $D, D', \ldots$ (also $C = C', \ldots$ and $B = B' \ldots$). This ensures that the figure is joined up as a torus:

(3b) $A = D$

From (3a) and (3b) we deduce:

(3) C is closer in the z-direction than A.

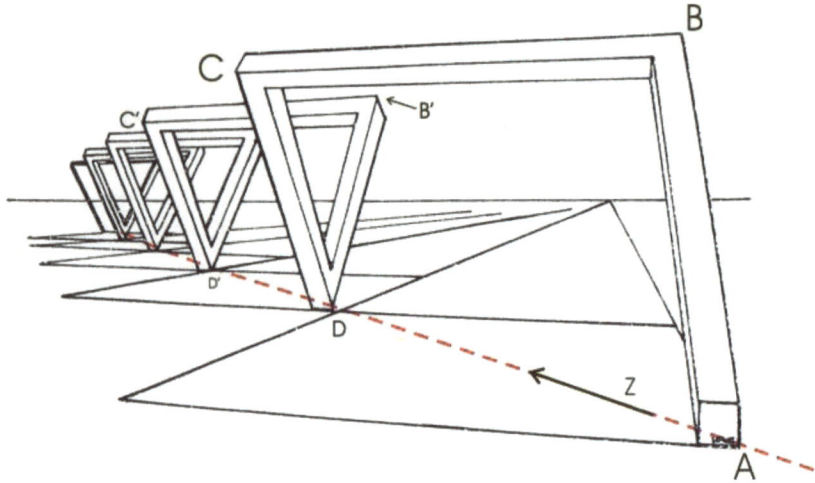

Figure 13.11: The impossible triangle unrolled

We then proceed with the deduction of the contradiction as on p.121, using the transitivity and irreflexivity of *closer*.

There are two consistent mathematical theories in this story. One theory T_1 describes what we seem to see, a single triangle/torus, but locates it in $R^2 \times S^1$. The other theory T_2 contains our expectation, that the space we live in is infinite in all directions, that is R^3, which in particular has a z-direction which is infinite. Strictly speaking in the latter theory the further spirals going off to infinity as Francis drew them are not necessary: a single spiral containing the points A and D is sufficient.

In T_1, $A = D$, since the figure is a torus (joined-up). In T_2, $A \neq D$, since by transitivity A is closer in the z-direction than D. The proposed inconsistent theory **T** is defined to be $Merge(T_1, T_2)$ which is the deductive closure of $T_1 \cup T_2$. In **T**, we have (5) above (p.121), namely that A is closer in the z-direction than A, since we have in T_2 that A is closer in the z-direction than D, and in T_1 that $A = D$. In **T**, we also have (6) above (P.121), namely that A is not closer in the z-direction than A, since this holds in both T_1 and T_2. This would seem then to be a complete explanation of our cognitive intuitions about the triangle. In a nutshell:

Theorem 87 *The theory of the inconsistent triangle* **T** *is given by* $Merge(T_1, T_2)$, *where* T_1, T_2 *and* $Merge$ *are as described above.*

13.6 Conclusion

There are various red herrings to avoid, as they do not get to the essence of the matter. Sometimes it is thought that the triangle has three 90° corners, which no triangle has, so that we should be looking for a geometry which permits triangles with three 90° corners. There are such geometries, of course, for example certain triplets of intersecting great circles on the surface of the sphere. But the four-sided figures that Thaddeus Cowan drew equally have 90° corners, and their four corners add up to the usual 360°, so that their inconsistency cannot be accounted for in this way. In any case, I for one do not see the triangle as having right-angled corners, they look like the regulation 60° to me.

Perspective is similarly irrelevant. The iconic triangle above does not contain perspective. There is, however, a subtle point which supports the present analysis. Penrose added perspective to the triangle, and there is no doubt that this enhances the effect. But why? *Because the perspective enhances the impression that the sides are dipping into the page, away from the plane of the paper.* This is exactly the point that Cowan's experiments, and the above account, turn on.

It is sometimes suggested that the key to understanding the triangle is the principle that objects with zero angular separation look joined (Gregory 1997). Now it is certainly true that photographs have been taken of objects that look like the triangle but which are in reality not joined up to form toruses. The open figure drawn by Francis (Figure 13.11), photographed along the z-axis, would be one such. But this cannot be the whole story. At most, it serves to explain why the content of our cognitive state contains the proposition that $A = D$. This is part of the explanation, as we have seen, but it is not the whole explanation since no contradiction has been shown to be generated. In a similar vein, we might have stopped our analysis above with the consistent theory of the triangle in $R^2 \times S^1$, as Francis did. This theory also contains $A = D$, but we would have been left with a gap in our understanding, namely the content of how exactly the triangle looks, when it looks impossible.

Finally, note the contribution of Section 13.3 to the *classification problem*. The two triangles are obviously very similar, but the differences in the proof of inconsistency indicate that they are different on a sufficiently fine-grained scheme of classiication.

In sum, the important thing to bear in mind is that any account which omits the inconsistent content of these theories is omitting to explain the strong sense we have of the impossibility of the objects they depict.

Figure 13.12: Two to three to two

Chapter 14

The Stairs

14.1 Introduction

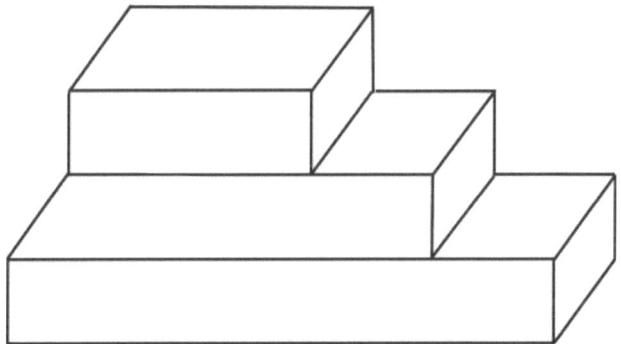

Figure 14.1: The stairs

In this chapter, we will investigate the Stairs, Figure 14.1. As in previous analyses, we first find a logical description according to which it is a paradox. It does not seem to be an occlusion paradox, however, if only because there appears to be nothing that is hidden from sight, nothing to be revealed by changing occlusions. Accordingly, no argument that it is an occlusion paradox is offered. The logical description does however make for a reasonable cognitive story, again in line with the leading epistemic motivation in all these inconsistent images so far. We then see that the ensuing inconsistent mathematical theory is different from those encountered to date, in using the notion of an anti-heap. As usual, for the argument to proceed, the stairs figure needs some labels, Figure 14.2.

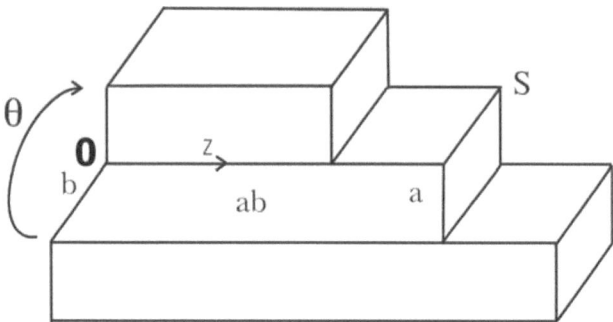

Figure 14.2: The stairs labelled

14.2 Logic

Considering the two lines a and b, we have:

(1) a is vertical

(2) b is horizontal.

In affirming these propositions, notice is taken of the role of the complex of stairs S at the right-hand end of the diagram. Even with S covered over, the line a would still be up the page and therefore arguably vertical; but with S in place, a becomes a riser in the stair complex S, which reinforces the vertical look of a, that is (1). In addition, the other risers in S are parallel to the two risers on the stairs at the left hand end of the diagram, and their treads are parallel to the line b, which imply (2).

We also have:

(3) a is coplanar with b

This is essential for the paradox, and is justified as a description of how the surface ab containing a and b looks. Also essential is:

(4) a and b arbitrarily extended do not intersect.

This looks correct in terms of the diagram; and it is certainly necessary, for without it all of (1), (2) and (3) are consistent with each other. However, taking (3) and (4) together, we have that

(5) a and b are parallel.

Applying this to (1), we have:

(6) b is vertical.

The contradiction is then generated by noting:

(7) vertical $\rightarrow \neg$ horizontal.

In passing, applying (5) to (2) also gives:

(8) a is horizontal

which similarly yields a paradox.

14.3 Cognition

This is a fair representation of the look of the stairs, I contend. However, particular attention should be paid to premiss (3), coplanarity. As with the case of the triangle, it is possible to have a consistent structure looking like the stairs, namely one with a *twisted* surface joining a to b, not a plane. It is therefore also possible to build such an object, with the caution that the twisted surface should have a matt texture, so that the twist is not revealed by the effects of shine. Without such effects, it is impossible to see the twist. Here the cognition has a *default setting*, as it had with the triangle. When the twist cannot be seen, the mind defaults to *flat*, hence coplanar. You can see the stairs as twisted, just as you can see the triangle as a figure which is not-joined-up; but the perceptual module naturally outputs a simpler default story in each case, which serves to explain the paradoxical appearances.

14.4 Mathematics

As we have seen, to find an inconsistent mathematical theory, it is useful to separate out consistent but mutually-incompatible theories and then merge them. In the present case, the simplest way to do this is to describe it as having a twisted surface, which supports the differing numbers of stairs at each end; and then add sufficient identifications for it to be a plane, which generates the contradiction.

For the ab surface to be (consistently) twisted, the best description uses *cylindrical coordinates*. These are triplets (r, θ, z) where (r, θ) are polar coordinates and z is an additional Cartesian z−axis. One can fix the origin of these coordinates at O, The line b is at $\theta = 0$, and occupies values for its r-coordinate ranging from 0 to l_b, the length of b (which is also the length of a). The z-axis is down the page, from left to right. As z increases, θ rotates downward, decreasing from 0 to $-\pi/2$. At $\theta = -\pi/2$ and $z = l_{ab}$, the length from a to b, the surface coincides with the vertical line a. Thus the twisted surface ab_{twist} is:

(9) $ab_{twist} = \{(r, \theta, z) : 0 \le r \le l_b \ \& \ 0 \le z \le l_{ab} \ \& \ -\pi/2 \le \theta \le 0 \ \& \ \theta = kz$ where k is $-\pi/2l_{ab}\}$.

The condition $\theta = kz$ ensures a single value of θ for a single z, varying linearly as z increases from left to right. At fixed θ and z, all points with r-coordinates between 0 and l_b are on the surface.

Since $a = \{(r, kl_{ab}, l_{ab}\}$ and $b = \{(r, 0, 0)\}$, we also have:

(10) a and b are wholly included in ab_{twist}.

Now, for a flat plane ab_{flat}, we can take the flat plane starting at b, extending horizontally down the z-axis, ending by meeting a at right angles. That is:

(11) $ab_{flat} = \{(r, 0, z) : r$ and z are as in (9)$\}$

Since having the same θ-coordinate suffices for being a plane, we also conclude that:

(12) ab_{flat} is (part of) a plane.

Therefore, it suffices for a and b to be coplanar, if ab_{twist} is identical with ab_{flat}. As we have seen, to include both theories it is necessary to include into the larger theory all denials of atomic sentences as well, which will make the theory inconsistent. The most direct way to ensure this, is with an (inconsistent) *anti-heap*. An anti-heap identifies all the points in between two given points (see Mortensen 2002). Strictly speaking, we require a collection of anti-heaps, one for each value of z (in other words a fibration). Primitive relations of the theories are $\{=, \in\}$, and the background logic as usual is 3-valued closed set logic. Then we set:

(13) a) If (r, θ, z) is the same point classically as (r', θ', z') then $I((r, \theta, z) = (r', \theta', z')) = T$; else

b) if $kz \le \theta, \theta' \le 0$ then $I((r, \theta, z) = (r, \theta', z)) = B$; else

c) $I((r, \theta, z) = (r', \theta', z')) = F$.

14.4. MATHEMATICS

The condition (b) ensures that points having the same (r, z)−coordinates are identified if their θ−coordinates fall within the bounds from kz to 0, the bounds given by ab_{twist} and ab_{flat} respectively. Note that this is an identification of points, not their real number coordinates. That is, arithmetical relations between the r−coordinates, or between the z−coordinates, are unaffected by identifications between points having differing θ−coordinates.

Also, for set membership:

(14)   a) If (r, θ, z) belongs classically to the set S then $I((r, \theta, z) \in S) = T$; else

b) if $(r, \theta, z) = (r', \theta', z')$ holds and $(r', \theta', z') \in S$ holds, then $I((r, \theta, z) \in S) = B$; else

c) $I((r, \theta, z) \in S) = F$.

And, for set inclusion:

(15)   a) If S_1 is classically included in S_2, then $I(S_1 \subseteq S_2) = T$; else

b) if for every $(r, \theta, z) \in S_1$ that holds there is an (r', θ', z') such that $(r, \theta, z) = (r', \theta', z')$ holds and $(r', \theta', z') \in S_2$ holds, then $I(S_1 \subseteq S_2) = B$; else

c) $I(S_1 \subseteq S_2) = F$.

And finally, for set identity:

(16)   a) If S_1 is classically identical with S_2, then $I(S_1 = S_2) = T$; else

b) if $S_1 \subseteq S_2$ and $S_2 \subseteq S_1$ both hold, then $I(S_1 = S_2) = B$; else

c) $I(S_1 = S_2) = F$.

With these assignments, it is clear that:

(17) $ab_{twist} \subseteq ab_{flat}$ and $ab_{flat} \subseteq ab_{twist}$ both hold. So:

(18) $ab_{twist} = ab_{flat}$ holds.

But now we can complete our argument, in the manner indicated in Section 14.2 above. The mathematical model validates a key premiss. Thus, both the lines a and b are included in ab_{flat}, and ab_{flat} is a plane, so a and b are coplanar, which is (3) above (p. 133). The rest follows in the fashion set out there.

14.5 Conclusion

Making the twisted plane the same as the flat plane is what makes the logical argument of the first section work. We postulated that the default setting for our perceptual cognitive apparatus is that the plane ab is flat. But by the construction in the preceding section, we can make this come out as holding (inconsistently) in an appropriate mathematical theory. Thus the mathematics implies the cognition, the cognition motivates the mathematics, and both support the key premiss of the logical argument.

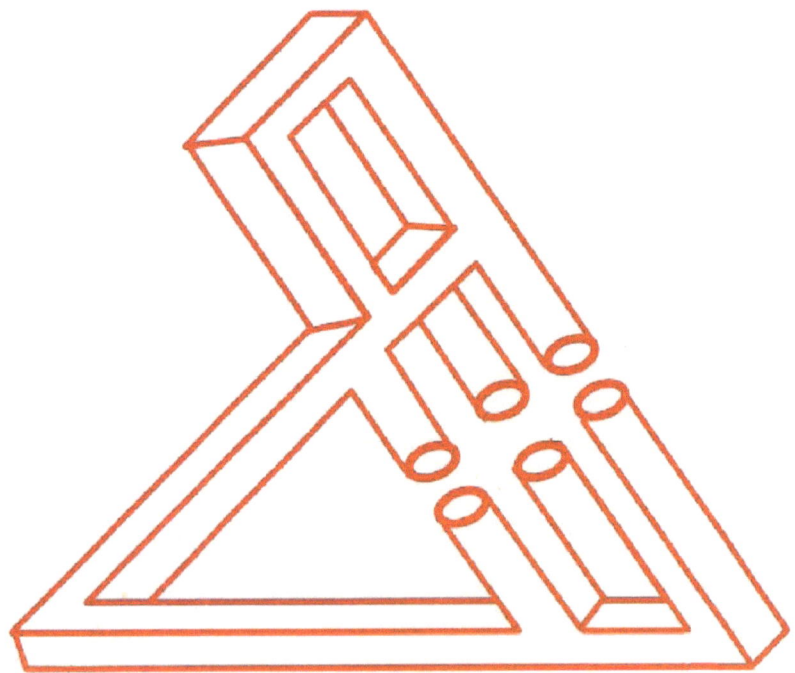

Figure 14.3: Schuster device

Chapter 15

The Fork

15.1 Introduction

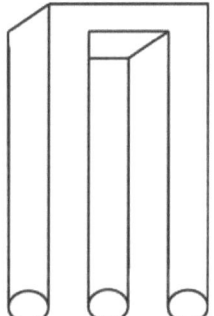

Figure 15.1: The fork

We come to the final class of images that we will be discussing, exemplified by Schuster's fork, Figure 15.1 The fork here has a top like a Box, a bottom like Pipes, and the aspect of impossibility. The style of analysis will be familiar from the preceding chapters. First we have a logical analysis, to demonstrate a contradiction and thus the existence of a paradox. As in the previous chapter, no attempt is made to show that it is an occlusion paradox. While the fork appears to harbour occlusions, there seems to be no reason for thinking that they contribute to the effect. Then we identify cognitive content contributing to the paradox, which explains why the content of the look is paradoxical. Finally, a mathematical theory is defined that implies the content. As usual, labels are necessary (see Fig 15.2)

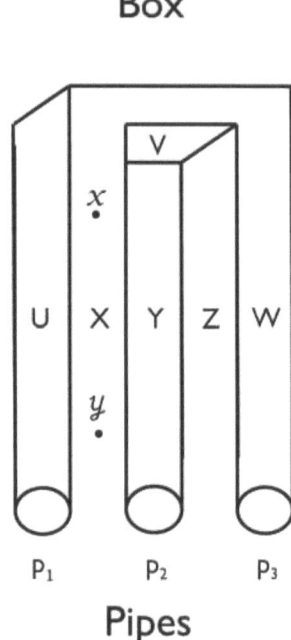

Figure 15.2: The fork labelled

15.2 Logic

The image on the page is a 2-D image of an 3-D object, in the sense that behind a point on the page is either the surface of the object, which is a 3-D material surface, or else there is no matter behind the point, only empty space unoccupied by matter. The construction of the image suggests to the cognition which points on the page have matter behind them, and which do not. The cognition then forms an hypothesis about what sort of 3-D object could have a surface that projects onto the page with that image.

For example, the construction of the top half, the Box, suggests that the point x has matter behind it. This is strengthened if the bottom half of the figure is covered. The appearance of Box also suggests that the part containing the point x is occluding parts of the box behind it, which is only possible if there is matter at x or behind it. So, writing "Mx" for "there is matter behind x", we have:

(1) Mx

15.3. COGNITION

In contrast, the construction of the bottom half, the Pipes, suggests that the point y has no matter behind it:

(2) $\neg My$

Now the role of the edges in an image of a 3-D object, is twofold: (a) to separate regions behind which there is matter, from regions behind which there is no matter, and (b) to indicate a corner or fold in matter. What edges do not do, is to separate empty regions from empty regions. That is, a region such as X which is not crossed by any edges, is such that if there is matter behind some point in X, there is matter behind all points in X, and if there is no matter behind some point, there is no matter behind any point. We might call this a *uniformity* premiss. That is:

(3) $Mx \leftrightarrow My$, and:

(4) $\neg Mx \leftrightarrow \neg My$

The propositions (1)-(4) are mutually contradictory.

There are other potential contradictions that might be squeezed out of this image. For instance, the region Y is similarly not crossed by edges, and so one might identify a matter-occupied point near the bottom pipe, and a matter-empty point at the top near the Box, to generate the contradiction. This has some plausibility, but has to deal with the objection that Y may be seen as wholly matter-occupied (separated from the rest of the Box by the fold at the top). Again, the region Z is not crossed by edges, clearly without matter at the bottom near the pipes, and arguably with matter at the top near the Box. But it has to deal with the objection that the region may be seen as simply empty, that matter does not begin until the other side of the top edge, the region above Y. This is perhaps harder to sustain, because the Box end of Z looks like it has a diagonal fold, which would put matter below that. At any rate, the argument (1)-(4) above appears to be the most defensible.

15.3 Cognition

The obvious premisses to focus on are (3) and (4), the uniformity premisses. Now there is a way in which uniformity could be false and not have that marked by a visible edge. This is, that matter *attenuates* between x and y, so that there is no definite boundary between matter and no-matter. So, for a cognitive reinforcement of the uniformity premiss one would need to find reason to deny attenuation. One point here is that the region X is uniformly empty of colour. Contrast this with when one attempts to add colour to the fork image: it is tempting to make the colour of the empty region around y shade into more Box-like tones around x. This takes away the inconsistent effect. As Bruno Ernst observes, the fork "cannot be coloured in" (Ernst 1986, 80). So the uniform white in X, and the absence of an edge inside it, suggest that the uniformity premiss is a default setting of the cognition.

This is a fair point, and it can be made stronger. No mention has been made so far of the conditions of observation of the image. One starts from one end, say the Pipes. Cover over the Box (or simply ignore it). There is no disputing what is matter and what is not, and that there is no matter at the point y. The Pipes with the Box covered are consistent. Moving the eye and the cover upward, it remains Pipes. This is a reasonable cognitive default mechanism having a sorites character: if you start with Pipes and no-matter, vary some parameter continuously, with no discernable change other than an irrelevant lengthening of the Pipes, then you stay as Pipes and no-matter. I stress that this is a cognitive mechanism, I am not proposing that it is what happens in existence (otherwise the sorites fallacy would not be fallacious, there would be no attentuation, and bald would be hairy). Rather, what I am aiming to describe is how there looks to be an incongruity when the eye eventually alights on the uncovered Box. Non-matter in the region X is carried upward until it encounters a point such as x where, by virtue of the nearby presence of the Box, there is indisputably matter. By the look of things, then, x should be both M and $\neg M$.

The uniform look spreads downward if one repeats the exercise starting at the top Box end, where the point x is M. Confronted with a point y where there is clearly non-matter, one similarly concludes that y is both M and $\neg M$.

This argument might seem to justify the cognitive contradiction directly, without the assistance of the uniformity premiss. But really it supports uniformity at the same time. As the eye moves downwards, the points one encounters remain the same as they started, namely M, because the mind is given no reason to vary its judgement, Box. Similarly when the gaze moves upwards. Uniformity, so to speak, is the expression of this lack of cognitive variation.

15.4 Mathematics

The points just made imply that any inconsistent mathematical theory expressing these ideas should have $M \& \neg M$ contradictions at both x and

15.4. MATHEMATICS

y, and all the points in between, since the moving mind carries the M and the $\neg M$ all the way in opposite directions. The problem, as usual, is to find consistent sub-theories which merge together into the inconsistency, always noting that these must be the sub-theories of 3-D objects. It is sufficient to describe the surface of matter behind the image, as material points inside the object cannot be seen from any angle.

Let us take stock of what we must have, and what is optional.

(5) The space outside both the Box and the Pipes is consistently $\neg M$.

The construction of the Box implies that :

(6) The points in the regions U and V are consistently M.

The Pipes agree with (6) on U; and either agree with (6) on V, or at least do not disagree on V. The Pipes similarly have consistent bottom ends P_1, P_2, P_3.

(7) The points in the regions P_1, P_2, P_3 are consistently M.

Now, from our analysis above:

(8) The points in the region X are inconsistently $M \& \neg M$.

But what of Y and Z? We saw in the first section that there are options. To simplify, I will treat these as contradictory, because of (a) the above argument for their inconsistency similar to that for X (see under (4)), and (b) because of their proximity to X. The reasonable place to draw the line for inconsistency, is to declare inconsistent the region between the bottom of the Box, and the top of the Pipes. Moving upward, the influence of the Pipes extends upward to the bottom of the Box. Moving downward, the influence of the Box extends downward to the top of the Pipes. After that, the part which is nearby, totally dominates. So:

(9) The points in the regions Y and Z are inconsistently $M \& \neg M$.

Now we note that both the Box interpretation and the Pipes interpretation agree that the region W is M. However, this raises the issue of whether to draw a line between the M in W and the contradictions in X, making the former consistent. The simplest solution is to allow the contradictions to spread throughout the XW region, which has the advantage that uniformity is preserved in that region as well. In support, it must be conceded that phenomenologically the XW region looks odd, open at one end and not at the other. Hence:

(10) The points in XW are inconsistently $M \& \neg M$.

These decisions allow at least one inconsistent theory to be defined. Let B be a (consistent) 3-D surface having the aspect (projection) of the Box image stretching down to the top of the Pipes, and let P be a 3-D surface with the aspect of the Pipes image, stretching up to the bottom of the Box. Let $proj(x)$ be the projection of a point x from either surface, onto the plane of the paper. We distinguish the material status of points x behind the plane of the paper by means of the predicate $M@x$. Thus $My \leftrightarrow (\exists x)(M@x \ \& \ proj(x) = y)$. Then we can define an interpretation I:

(11) a) If $x \in B \cup P$ and $proj(x) \in U \cup V \cup P_1 \cup P_2 \cup P_3$ then $I(M@x) = T$; else

b) if $x \in B \cup P$ and $proj(x) \in X \cup Y \cup Z \cup W$ then $I(M@x) = B$; else

c) $I(M@x) = F$.

The interpretation is extended to non-atomic sentences in the usual way. Then define the associated theory of the fork Th_{fork} by:

(12) $Th_{fork} := \{A : I(A) \in \{T, B\}\}$.

Summing up:

Theorem 88 *The theory of the fork Th_{fork} is inconsistent. It satisfies the conditions (1)-(10), including uniformity, so that the logical argument (1)-(4) for the paradoxicality of the fork is correct according to the theory.*

15.5 Conclusion

This completes our discussion of The Fork, and indeed the book. It is to be hoped that the reader comes away with an expanded view of what is possible for paraconsistency. Not all of the book has been about visual paradoxes. The first part was not, concentrating instead on ways of generating inconsistent theories deriving from topological operations, and from groups. This enabled various traditional geometrical topics to be adapted as applications of the theory of inconsistency. Visual paradoxes were then addressed in the second part. A style of analysis of visual paradoxes was proposed, which finds logic, cognition and mathematical theories to be mutually supporting. This enables rigorous proofs of paradoxicality, and indeed occlusion-paradoxicality where appropriate. The

15.5. CONCLUSION

leading hypothesis has been that inconsistent images have a cognitive explanation, but conversely they are a source of inconsistent mathematical theories from which cognitive intuitions can be extracted. It was also claimed earlier that impossible images come in four basic kinds. Arguing this adequately requires some sort of framework of essential features for an inconsistent description. To date, we had only an intuitive survey in support of this hypothesis. The present methodology recommends itself as permitting further rigorous proofs along these lines.

Figure 15.3: The watcher

Bibliography

[1] Agoston, Max, (1976), *Algebraic Topology, A First Course*, New York, Marcel Dekker.

[2] Anderson, A, and Belnap, N (1975), *Entailment* , Princeton, Princeton University Press.

[3] Birkhoff, G., and Mac Lane, S., (1965), *A Survey of Modern Algebra, (3rd edn.)*, new York, Macmillan.

[4] Chang, C.C. (1958), "Algebraic Analysis of Many-Valued Logic", *Transactions of the American Mathematical Society*, 88, 467-490.

[5] —, (1963), "A Logic with Positive and Negative Truth Values", *Acta Philosophica Fennica*, 16, 19-39.

[6] Cignoli, R, D'Ottaviano, I, and Mundici, D, (2000), *Algebraic Foundations of Many-valued Reasoning,* Dordrecht, Kluwer.

[7] Cowan, Thaddeus (1974), "The Theory of Braids and the Analysis of Impossible Pictures", *Journal of Mathematical Psychology*, 11, 190-212.

[8] —, (1977a), "Supplementary Report: Braids, Side Segments and Impossible Figures", *Journal of Mathematical Psychology*, 11, 254-260.

[9] —, (1977b), "Organising the Properties of Impossible Figures", *Perception,* 6, 41-56.

[10] — and R.Pringle, (1978), "An Investigation of the Cues Responsible for Figure Impossibility", *Journal of Experimental Psychology: Human Perception and Performance*, 4, 112-120.

[11] Coxeter, H.S.M, (1961), *Introduction to Geometry*, Wiley.

[12] Del Prete, Sandro, (2008), *Master of Illusions*, New York, Sterling Publishing.

BIBLIOGRAPHY 143

[13] Draper, S.W., (1978), "The Penrose Triangle and a Family of Related Figures", *Perception,* 7, (283-296).

[14] Dunn, J.M. (1979), "A Theorem in Three Valued Model Theory, with Connections to Number Theory, Type Theory and Relevant Logic", *Studia Logica,* 38, 149-169.

[15] Ernst, Bruno, (1986), *The Eye Beguiled,* Cologne, Taschen.

[16] Francis, G.A.(1987), *A Topological Picture Book,* Springer-Verlag.

[17] Galli, A., Lewin, R., and Sagastume, M., (2004), "The Logic of Equilibrium and Abelian lattice Ordered Groups", *Arch.Math.Log.,* 43, 141-158.

[18] Greenberg, M, and Harper, J, (1981), *Algebraic Topology, A First Course,* California, Addison-Wesley.

[19] Gregory, Richard L., (1970), *The Intelligent Eye,* London, Weidenfeld and Nicholson.

[20] ——, (1997), "Knowledge in Perception and Illusion", *Philosophical Transactions of the Royal Society of London* B, 352, 1121-1128.

[21] Hatcher, Allen, (2002) *Algebraic Topology,* Cambridge, Cambridge University Press.

[22] Kelley, J.L. (1955), *General Topology,* New York, Van Nostrand.

[23] Kim, Scott, (1978), "An Impossible Four-dimensional Illusion", in D.Brisson (ed), *Hypergraphics,* 186-239.

[24] Lewin, R, and Sagastume, M, (2002), "Paraconsistency in Chang's Logic with Positive and Negative Truth Values", in Carnielli, W., *(et. al. eds.), Paraconsistency, the Logical Way to the Inconsistent,* New York, Marcel Dekker, 381-396.

[25] Meyer, R.K, and Slaney, J, (1989), "Abelian Logic from A to Z", in G.Priest *(et. al. eds), Paraconsistent Logic, Essays on the Inconsistent,* Munchen, Analytica, Philosophia Verlag, 245-288.

[26] —-, "A, Still Adorable", (2002), in Carnielli, W. *(et. al. eds.) Paraconsistency, The Logical Way to the Inconsistent,* New York, Marcel Dekker, 241-260.

[27] Mortensen, Chris, (1995), *Inconsistent Mathematics,* Dordrecht, Kluwer Mathematics and Its Applications Series.

[28] —- (1997), "Peeking at the Impossible", *Notre Dame Journal of Formal Logic,* 38, 527-534.

[29] —- (2002), "Towards a Mathematics of Impossible Pictures", in W.Carnielli *(et al eds), Paraconsistency, the Logical Way to the Inconsistent*, New York, Marcel Dekker, 445-454.

[30] —- (2006), "An Analysis of Inconsistent and Incomplete Necker Cubes". *Australasian Journal of Logic,* 4, 216-225.

[31] —- (2009), "Linear Algebra Representation of Necker Cubes 2: The Routley Functor and Necker Chains", *Australasian Journal of Logic,* 7, 11-26.

[32] —- and Leishman, S, (2009), "Linear Algebra Representation of Necker Cubes 1: The Crazy Crate", *Australasian Journal of Logic,* 7, 1-10.

[33] —- and Quigley, P, (2005), "Cubic Logic, Ulam Games and Paraconsistency", *Journal of Applied Nonclassical Logic,* 15, 59-69.

[34] —-, Leishman, S, Quigley, P and Mercier J, Inconsistent Images website, http://www.hss.adelaide.edu.au/philosophy/inconsistent-images/

[35] Mundici, D., (2002), "Fault Tolerance and Rota-Metropolis Cubic Logic", in Carnielli, W., *(et al), Paraconsistency, the Logical Way to the Inconsistent,* New York, Marcel Dekker, 397-409

[36] Penrose, L.S. and Roger, (1958), "Impossible Objects: A Special Kind of Illusion", *British Journal of Psychology*, 38, 49

[37] Penrose, Roger, (1991), "On the Cohomology of Impossible Pictures", *Structural Topology*, 17, 11-16.

[38] Priest, Graham, (2006), *In Contradiction,* second ed, Oxford, Clarendon.

[39] Priest, Graham, (2001), *An Introduction to Non-classical Logic,* Cambridge, Cambridge University Press.

[40] Rasiowa, Helena and Sikorsky, Roman, (1968), *The Mathematics of Metamathematics,* (2 ed revised), Warsaw, Polish Scientific Publishers.

[41] Rota, G-C, and Metropolis, N, (1978), "Combinatorial Structure of the Faces of the n-Cube", *Journal of Applied Mathematics,* 35, 689-694.

BIBLIOGRAPHY

[42] Routley, R and V, (1972), "The Semantics of First Degree Entailment", *Nous,* 6, 335-359.

[43] Seckel, Al, (2004), *Masters of Deception,* New York, Sterling Publications.

[44] Shin, Sun-Joo and Lemon, Oliver, (2003), "Diagrams", *Stanford Encyclopedia of Philosophy,* (Winter 2003 edition), http://plato.stanford.edu.

[45] Simmons, G., (1963), *Introduction to Topology and Modern Analysis,* McGraw-Hill.

[46] Terouanne, E., (1983a), "On a Class of Impossible Figures: a New Language for a New Analysis", *Journal of Mathematical Psychology,* vol 22, No 1, 24-47.

[47] —-, (1983b), "Impossible Figures and Interpretations of Polyhedral Figures", *Journal of Mathematical Psychology,* 27, 370-405.

[48] Thro, E.B., (1983), "Distinguishing Two Classes of Impossible Objects", *Perception,* vol 12, No 6, 733-751.

[49] Whiteley, W. (1979), "Realizability of Polyhedra", *Structural Topology,* 1, Montreal, U. of Montreal Press.

Appendix A

Gallery of Inconsistent Images

Figure A.1: Square triangle

Figure A.2: Stairs and fork

Figure A.3: Polypod

148 APPENDIX A. GALLERY OF INCONSISTENT IMAGES

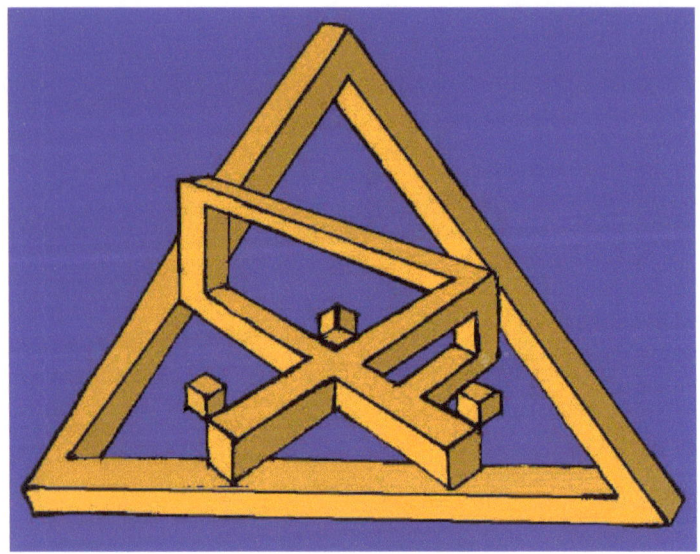

Figure A.4: Gemeinschaft

Figure A.5: Entangled

Figure A.6: Well-supported

Figure A.7: Open cage

150 APPENDIX A. GALLERY OF INCONSISTENT IMAGES

Figure A.8: Plane

Figure A.9: Cochran crate

Figure A.10: Dis con nec ted

Figure A.11: Rack

152　　APPENDIX A. GALLERY OF INCONSISTENT IMAGES

Figure A.12: Intersecting frames

Figure A.13: Obscure camera

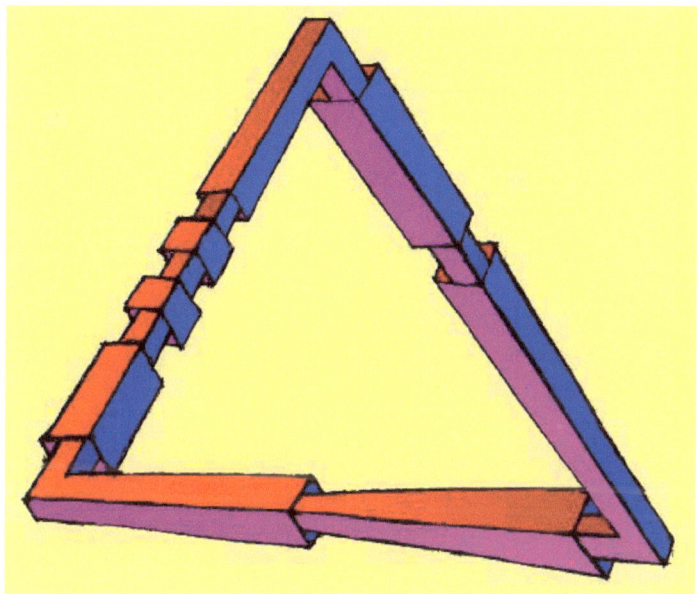

Figure A.14: Notches

Figure A.15: Not real straight

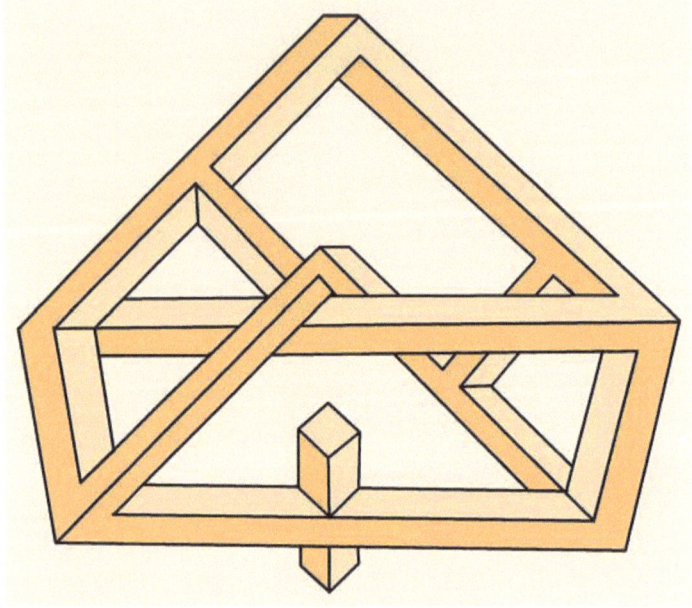

Figure A.16: Joined

Figure A.17: Superstructure in perspective

Figure A.18: Impossible Borromean rings

Appendix B

Mostly B&W Images

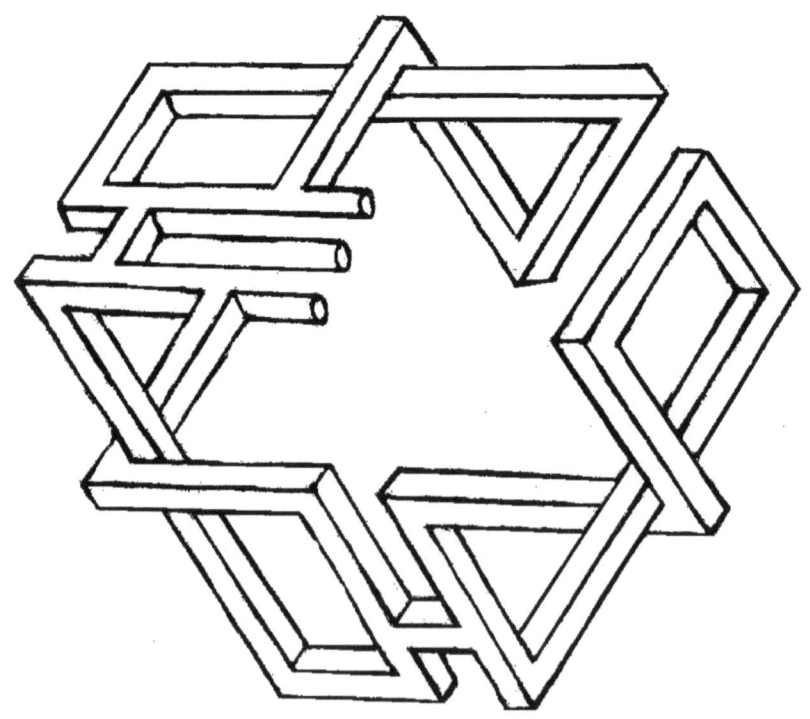

Figure B.1: Space-time grid

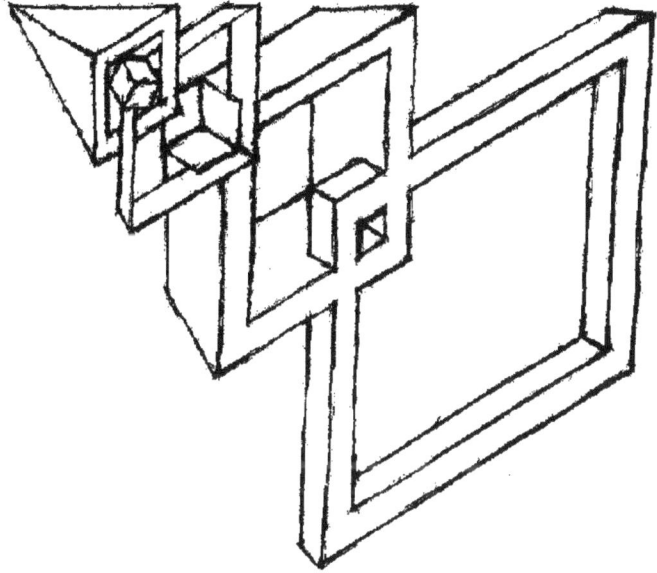

Figure B.2: Warp 10

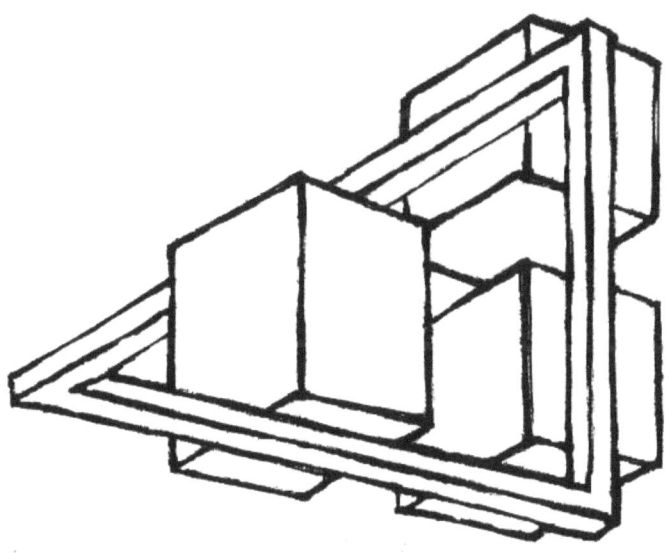

Figure B.3: Triangle with cubes

Figure B.4: Mrs Bates' mirror

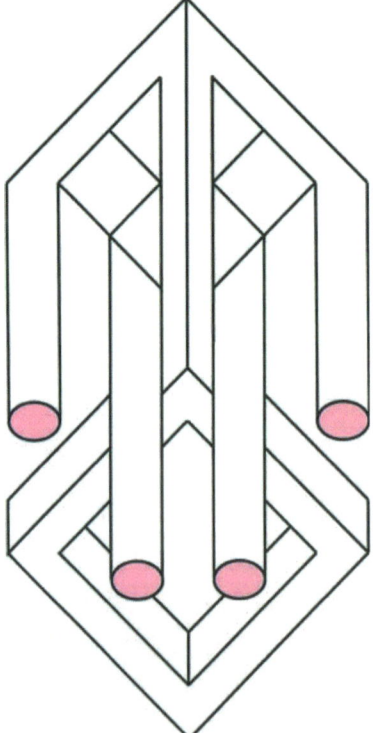

Figure B.5: Back to front

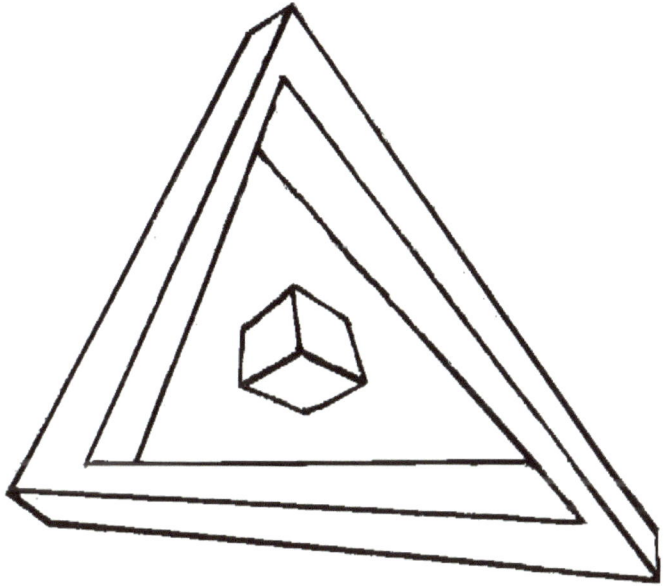

Figure B.6: Reverse perspective

Figure B.7: Bird in space (homage to Brancusi)

Index

algebra, 52
    set of generators, 52
algebras
    similar, 52
ambiguous images, 71
Anderson and Belnap, 27, 82
anti-diagonal matrix, Anti-Id, 93
Aristotle, 25
atomic extension, 16
atomically disjoint, 16
axiom of relativity, 30

Birkhoff and Mac Lane, 42
boundary function, 60
Brouwerian algebra, 8

Cayley, 21
chain complex, 61
Chang, 21, 26, 28
Cignoli, 21, 25, 33
closed complement, 8
closed set negation, 8
Cochran's cube
    the crazy crate, 80
coladd, 101
colswap, 101
consistent theory of rotational symmetries, 47
contravariant, 106
covariant, 106
Cowan, 4, 73, 130
Cowan and Pringle, 121, 126
Coxeter, 42

cycles, 60

de May, 73
degree of inconsistency (of Necker chain), 112
del Prete, 73
determinant, 94
diagonal matrix, Id, 93
dialetheism, 5
dihedral group, 49
discrete topology, 14
Double Negation, 48
Duchamp, 72
Dunn, 53

ECQ, 3
elementary operations on matrices, 101
Erlangen, 20
Ernst, ix, 72, 80, 140
Ernst stairs, 76
Escher, viii, 4, 73
Escher's cube, 79
Euclid, 20
Ex Contradictione Quodlibet, 3
exact sequence, 61
extendability lemma, 46
extension, 46

fission, 23
Francis, 4, 74, 128, 130
functional, 45
fusion, 23

INDEX

global inconsistency, 89
Gregory, 130
group, 20
    Abelian, 20
    absolute Routley star on, 63
    additive, 20
    homology, 61
    lattice ordered, 22
    multiplicative, 21
    quotient, factor, difference, 55
    relative Routley star on, 63
groups
    direct product, 50

Hasse diagram, 4, 24
Hausdorff space, 14
Heyting algebra, 7
higher order inconsistency, 109
homologous, 61
homomorphism, 52
    kernel of , 54
Humberstone, 88

identity and disidentity, conditions on, 17
impossible pictures, 71
in front of, 88
incomplete theories, 7
incomplete world, 7
inconsistent images, 71
    website, ix, 147
inconsistent theory, 5
inconsistent theory of rotational symmetries, 45
inconsistent world, 9
isometry, 42
isomorphism, 52

Kelley, 14
Klein, 20

lattice, 22
Leishman, ix, 91

Lewin-Sagastume, 27
local completeness, 85
local consistency, 85
logic, 3
    Abelian, 28
    closed set, 9
    cubic, 36
    intuitionist, 7
    modal, 5
    open set, 7
    relevant, 27
logical subtraction, 8
Lukasiewicz, 21
Lukasiewicz logic
    finite valued, 24
    infinite valued, 23

Magritte, 72
membership-theory, 15
Meyer, 28, 30
Mortensen, 18
Mundici, 40
MV-algebra, 22
    operations, 23

Necker chains, 108
Necker cube, 80
Newell, 72
normal topological space, 14
null space, 94
nullity, 94

occlusion, 86
occlusion paradox, 76
one-point theory, 15
open complement, 7
open set negation, 7

paraconsistency
    strong, 5
    weak, 5
paraconsistent, 3
parity, 126
Penrose, viii, 4, 73, 75, 130

Pinocchio negation, 40
Piranesi, 72
Plato, 4
Post-complete, 29
Priest, 46, 53
primary equation, 92
primary matrix, 92
prime, 72
projection homomorphism, 50

Quigley, ix, 32

Reutersvaard, viii, 4, 72
Rota-Metropolis, 36
Routley and Routley, 82
Routley functor, 47
Routley star, 48

Schuster, 138
Schuster fork, 76
scorched earth model, 18
Seckel, 73
secondary equation, 92
secondary matrix, 92
separation principles, 14
short exact sequence, 61
Simmons, 14
simplices, 58
simplicial complexes, 59
Slaney, 28, 30
subalgebra, 52
subgroup
    cosets of, 54
    normal, 54
switch, 102
symmetries of reflection, 49
symmetry, 42
symmetry group of rotations, 43

theory of model on a space, 13
topological complement, 10, 13
topological negation, 13
translational symmetry, 50
transparent, 45

transpose, 92
trivial, 7

Ulam games, 32
unit equation, 97

www.ingramcontent.com/pod-product-compliance
Lightning Source LLC
Chambersburg PA
CBHW041432300426
44117CB00001B/6